Scrapy网络爬虫开发实战

罗刚　编著

清华大学出版社
北京

内容简介

本书介绍如何学习和使用流行的Scrapy框架开发网络爬虫应用，主要内容使用Python开发网络爬虫，识别网页的编码，结构化信息的提取，Scrapy爬虫的示例使用，Scrapy Playwright抓取动态JS网站，将抓取的数据保存到数据库，部署、调度和运行Scrapy爬虫等。

本书适合作为高等院校计算机、软件工程专业本科生、研究生的参考书目，也适用于对Python网络爬虫领域感兴趣的人士参考阅读。

本书封面贴有清华大学出版社的防伪标签，无标签者不得销售。
版权所有，侵权必究。举报：010-62782989，beiqinquan@tup.tsinghua.edu.cn。

图书在版编目（CIP）数据

Scrapy网络爬虫开发实战/罗刚编著.—北京：清华大学出版社，2024.5
ISBN 978-7-302-65858-0

Ⅰ.①S… Ⅱ.①罗… Ⅲ.①软件工具－程序设计 Ⅳ.①TP311.561

中国国家版本馆CIP数据核字（2024）第060787号

责任编辑：张 敏 薛 阳
封面设计：郭二鹏
责任校对：胡伟民
责任印制：沈 露

出版发行：清华大学出版社
网　　址：https://www.tup.com.cn，https://www.wqxuetang.com
地　　址：北京清华大学学研大厦A座　邮　编：100084
社 总 机：010-83470000　邮　购：010-62786544
投稿与读者服务：010-62776969，c-service@tup.tsinghua.edu.cn
质 量 反 馈：010-62772015，zhiliang@tup.tsinghua.edu.cn
课 件 下 载：https://www.tup.com.cn，010-83470236
印 装 者：三河市君旺印务有限公司
经　　销：全国新华书店
开　　本：185mm×260mm　印　张：11.5　字　数：290千字
版　　次：2024年6月第1版　印　次：2024年6月第1次印刷
定　　价：69.80元

产品编号：104829-01

前言

Scrapy 是一个用 Python 语言编写的网络爬虫框架。它应用广泛——从数据挖掘到监控和自动化测试。它目前由 Zyte 公司维护，前身为 Scrapinyhub 公司，这是一家网络抓取和服务公司。本书介绍使用 Scrapy 开发网络爬虫应用。

本书共 8 章，第 1 章介绍网络爬虫开发快速入门；第 2 章介绍 Python 开发快速入门；第 3 章介绍使用 Python 开发网络爬虫；第 4 章介绍从互联网提取信息；第 5 章介绍使用 Scrapy 开发爬虫；第 6 章介绍分布式爬虫开发；第 7 章介绍如何开发网络爬虫用户界面；第 8 章介绍网络爬虫案例。

本书适合需要具体实现网络爬虫应用的开发人员或者对网络爬虫等相关领域感兴趣的人士参考，同时猎兔搜索技术团队已经开发出本书配套的培训课程和商业软件。

本书中的一些内容和现有的一些开源项目 Scrapy 等共同成长和发展。由于作者水平有限，书中疏漏之处在所难免，敬请广大读者朋友批评指正。

感谢早期合著者、合作伙伴、员工、学员、读者的支持，给我们提供了良好的工作基础，就像玻璃容器中的水培植物一样，这是一个持久可用的成长基础，技术的融合与创新无止境，欢迎一起探索。

目录

第1章 网络爬虫快速入门 ... 1
- 1.1 各种网络爬虫 ... 1
  - 1.1.1 通用爬虫 ... 1
  - 1.1.2 定向爬虫 ... 2
- 1.2 网络爬虫基本技术 ... 2
- 1.3 Windows 命令行 ... 3
- 1.4 上手 Scrapy 网络爬虫开发 ... 5
- 1.5 本章小结 ... 7

第2章 Python 开发快速入门 ... 8
- 2.1 变量 ... 8
- 2.2 注释 ... 8
- 2.3 简单数据类型 ... 9
  - 2.3.1 数值 ... 9
  - 2.3.2 字符串 ... 11
  - 2.3.3 数组 ... 14
- 2.4 字面值 ... 15
- 2.5 控制流 ... 15
  - 2.5.1 if 语句 ... 16
  - 2.5.2 循环 ... 17
- 2.6 列表 ... 17
- 2.7 元组 ... 21
- 2.8 集合 ... 22
- 2.9 字典 ... 23
- 2.10 函数 ... 23
- 2.11 模块 ... 26
- 2.12 检查字符串是否包含子字符串 ... 27

2.13	面向对象编程	28
2.14	泛型	30
2.15	日志记录	32
2.16	数据库	33
2.17	本章小结	35

第 3 章 使用 Python 开发网络爬虫 36

3.1	使用 BeautifulSoup 实现定向采集	36
3.2	URL 基础	39
	3.2.1 URI	40
	3.2.2 解析相对地址	40
	3.2.3 DNS 解析	41
3.3	网络爬虫抓取原理	42
3.4	爬虫架构	44
	3.4.1 基本架构	44
	3.4.2 分布式爬虫架构	46
	3.4.3 垂直爬虫架构	47
3.5	下载网页	48
	3.5.1 HTTP	48
	3.5.2 HTML 文档	53
	3.5.3 使用 curl 命令下载网络资源	53
	3.5.4 使用 wget 命令下载网页	55
	3.5.5 下载静态网页	56
	3.5.6 使用 Selenium 下载动态内容	59
	3.5.7 重试	61
3.6	下载图片	63
3.7	网络爬虫的遍历与实现	63
3.8	robots 协议	66
3.9	连接池	67
3.10	URL 地址查新	68
	3.10.1 Redis 数据库	69
	3.10.2 布隆过滤器	70
3.11	抓取 RSS	72
3.12	网页更新	72

- 3.13 进度条 ······ 74
- 3.14 垂直行业抓取 ······ 75
- 3.15 抓取限制的应对方法 ······ 76
  - 3.15.1 模拟浏览器访问 ······ 77
  - 3.15.2 使用代理 IP ······ 77
  - 3.15.3 抓取需要登录的网页 ······ 78
- 3.16 保存信息 ······ 79
  - 3.16.1 SQLite 数据库 ······ 79
  - 3.16.2 MySQL 数据库 ······ 80
  - 3.16.3 MongoDB 数据库 ······ 81
  - 3.16.4 存入 Elasticsearch 搜索引擎 ······ 84
- 3.17 本章小结 ······ 85

第 4 章 从互联网提取信息 ······ 86

- 4.1 识别网页的编码 ······ 86
  - 4.1.1 二进制流的编码 ······ 86
  - 4.1.2 识别编码的整体流程 ······ 87
- 4.2 正则表达式 ······ 88
- 4.3 结构化信息的提取 ······ 89
  - 4.3.1 解析 JSON ······ 89
  - 4.3.2 解析 XML ······ 90
  - 4.3.3 XML 接口 ······ 91
  - 4.3.4 lxml 处理网页 ······ 93
  - 4.3.5 使用 XPath 提取信息 ······ 93
  - 4.3.6 在 Chrome 浏览器中查找 Selenium WebDriver 的 XPath ······ 94
  - 4.3.7 CSS 选择器 ······ 95
  - 4.3.8 使用 Parsel ······ 95
  - 4.3.9 提取文本 ······ 97
  - 4.3.10 网页正文提取 ······ 97
- 4.4 从文件提取信息 ······ 99
- 4.5 本章小结 ······ 99

第 5 章 使用 Scrapy 开发爬虫 ······ 100

- 5.1 一个示例爬虫的演练 ······ 100
- 5.2 Scrapy Playwright 指南：渲染和抓取动态 JS 网站 ······ 101

5.3　将抓取的数据保存到 SQLite 数据库 ……………………… 108
5.4　将抓取的数据保存到 MySQL 数据库 ……………………… 114
5.5　将抓取的数据保存到 Postgres 数据库 ……………………… 120
5.6　Scrapyd：部署、调度和运行 Scrapy 爬虫 ………………… 126
5.7　Scrapy Cloud 托管爬虫 ………………………………………… 129
5.8　Twisted 框架 ……………………………………………………… 131
5.9　本章小结 ……………………………………………………… 133

第 6 章　分布式爬虫开发 ……………………………………… 134
6.1　简单的 Celery 任务 …………………………………………… 134
6.2　从任务进行分布式抓取 ………………………………………… 135
6.3　本章小结 ……………………………………………………… 141

第 7 章　开发网络爬虫用户界面 ……………………………… 142
7.1　Tkinter 简介 …………………………………………………… 142
7.2　网络爬虫图形用户界面 ………………………………………… 162
7.3　本章小结 ……………………………………………………… 165

第 8 章　案例分析 ……………………………………………… 166
8.1　影视采集器 …………………………………………………… 166
8.2　暗网爬虫 ……………………………………………………… 172
8.3　本章小结 ……………………………………………………… 173

第 1 章　网络爬虫快速入门

人工智能应用需要源源不断的数据来让应用跟上变化的世界。可以使用网络爬虫从互联网中自动获取数据。

在搜索引擎优化（SEO）领域，可以通过网络爬虫补充数据来实现更多的关键词收录。

网络爬虫经过最近几十年的快速发展，已经改变了人们获取信息的方式。从搜索引擎到推荐系统，都会用到网络爬虫技术。本书介绍采用流行的 Python 编程语言实现网络爬虫。

1.1　各种网络爬虫

有运行在大规模云计算平台的通用网络爬虫，还有一些行业垂直爬虫以及网站定向爬虫。通用网络爬虫是大鳄，每一只都有自己独立的领地。行业垂直爬虫是领头雁，是各行业的旗帜。而网站定向爬虫则像一只只小麻雀，麻雀虽小，五脏俱全。

1.1.1　通用爬虫

目前通用网络爬虫的组织方式主要有网络综合爬虫和网络主题资源爬虫两种。其中网络综合爬虫能够广泛地采集各互联网站点资源，并对其进行页面搜索，将索引结果存入索引数据库，供网络用户检索，并且能够提供互联网网络资源地导航功能的工具，如 Google、百度等。

Google、百度这样的公司需要大量的服务器和专业开发人员，运营开销大，如何在经济上可行就是一个问题。通用网络爬虫的主要收入是在搜索结果页中展示和用户输入的关键词相关的广告。条幅广告比关键词广告更早出现。按点击付费的关键词广告比条幅广告的收费额度低许多，点击一次广告可能只收几分钱，而条幅广告的计价单位至少在几百块。那些曾经被忽视的中小企业，一度被认为是游离在广告市场之外的客户，现在突然进入了互联网广告的生态系统。地球上最大的动物鲸鱼吃的是小鱼小虾，只有让更多的生物进入生态链，才能够产生庞大的顶级生物。

通用网络爬虫的企业是资本密集型企业，这样的公司往往前期有风险投资，有一定盈利后成为上市公司。

1.1.2 定向爬虫

垂直定向爬虫是针对某一个行业的专业爬虫,例如搜房(http://www.soufun.com/),39健康网上的搜索。垂直搜索是搜索引擎的细分和延伸,是对网页库中的某类专门的数据进行处理后再以某信息进行一次整合,定向分字段抽取出需要的数据进行处理后再以某种形式返回给用户。

垂直爬虫需要从茫茫的互联网中获取行业信息,信息按行业过滤和分类是必不可少的。垂直搜索引擎和普通的网页搜索引擎的一个最大区别是对网页信息进行结构化信息抽取,也就是将网页的非结构化数据抽取成特定的结构化信息数据,好比网页搜索是以网页为最小单位,基于视觉的网页块分析是以网页块为最小单位,而垂直搜索是以结构化数据为最小单位。然后将这些数据存储到数据库中,并进行进一步的加工处理,如去重、分类等。最后分词、索引再以搜索的方式满足用户的需求。

整个过程中,数据由非结构化数据抽取成结构化数据,经过深度加工处理后以非结构化的方式和结构化的方式返回给用户。

垂直爬虫的应用方向很多,比如企业库爬虫、供求信息爬虫、购物爬虫、房产爬虫、地理信息爬虫、音乐爬虫、图片爬虫……几乎各行各业各类信息都可以进一步细化成各类的垂直爬虫。

垂直爬虫的技术评估应从以下几点来判断。

(1)全面性:应该能从众多的来源采集信息。

(2)更新性:用户最好可以在几秒或几分钟内看到最新发布的信息。

(3)准确性:数据分类准确,不能包含重复冗余信息。

(4)功能性:功能完善,可以同时搜索文字信息、图片、视频、地理信息等。

1.2 网络爬虫基本技术

一个基本的爬虫包括采集数据下载器和运行状况监控面板等部分。

网络爬虫(Crawler)的主要目的是为获取互联网上的信息。网络爬虫利用主页中的超文本链接遍历Web,通过URL引用从一个HTML文档爬行到另一个HTML文档。http://dmoz.org是整个互联网抓取的入口。网络爬虫收集到的信息可有多种用途,如建立索引、HTML文件的验证、URL链接验证、获取更新信息、站点镜像等。网络爬虫建立的页面数据库,包含有根据页面内容生成的文摘,这是一个重要特色。

网站本身可以声明不想被网络爬虫抓取的内容。可以有两种方式实现:第一种方式是在站点增加一个纯文本文件,例如 http://www.baidu.com/robots.txt;另外一种方式是直接在 HTML 页面中使用 robots 的 meta 标签。在抓取网页时大部分网络爬虫会遵循 robot.txt 协议。

1.3 Windows 命令行

为了提高学习和工作效率,需要学会使用一些有用的软件工具,这里先介绍 Windows 命令行的使用。

假设有一个标准件工厂,在车间生产产品,在工地使用这些产品,类似地,往往在集成开发环境中开发软件。如果在 Windows 操作系统中运行开发的软件,则往往通过 Windows 命令行来运行。

在图形化用户界面出现之前,人们就是用命令行来操作计算机的。Windows 命令行是通过 Windows 系统目录下的 cmd.exe 执行的。可以在开始菜单的运行窗口直接输入程序名,回车后运行这个程序。打开开始→运行,这样就会打开资源管理器中的运行程序窗口。或者使用快捷键——窗口键+R 键,打开运行程序窗口。总之,输入程序名 cmd 后单击确定,出现命令提示窗口。因为能够通过这个黑屏的窗口直接输入命令来控制计算机,所以也称为控制台窗口。

通常用扩展名来表示文件的类别,例如,exe 表示可执行文件。文件名称由文件名和扩展名组成。文件名和扩展名之间由小数点分隔,例如 calc.exe。

使用 rmdir 命令删除目录,例如删除 opencvsharp 目录:

```
D:\soft>rmdir /s opencvsharp
```

为了方便从命令行安装软件,可以先安装软件包管理工具软件 Chocolatey。使用如下命令安装 Chocolatey:

```
@"%SystemRoot%\System32\WindowsPowerShell\v1.0\powershell.exe" -NoProfile -InputFormat None -ExecutionPolicy Bypass -Command "iex ((New-Object System.Net.WebClient).DownloadString('https://chocolatey.org/install.ps1'))" && SET "PATH=%PATH%;%ALLUSERSPROFILE%\chocolatey\bin"
```

然后可以使用 choco 命令安装一些开发用的软件,例如安装 git:

```
>choco install git
```

当我们建立或修改一个文件时,必须向 Windows 指明这个文件的位置。文件的位置由三部分组成:驱动器、文件所在路径和文件名。路径是由一系列路径名组成的,这些路径名之间用"\"分开,例如:

```
C:\Windows\System32\calc.exe
```

开始的路径往往是 C:\Users\Administrator。公园的地图上往往会标出游客的当前位置，Windows 命令行也有个当前目录的概念，这个 C:\Users\Administrator 就是当前路径。

可以用 cd 命令改变当前路径，例如改变到 C:\Windows\System32 路径。

```
C:\Users\Administrator>cd C:\Windows\System32
```

如果写 cd d:，这样的效果是改变当前路径到 d: 子目录。所以切换盘符不能使用 cd 命令，而是直接输入盘符的名称，例如想要切换到 d 盘，可以使用如下命令：

```
C:\Users\Administrator>d:
```

执行一个可执行文件：

```
C:\Users\Administrator>C:\Windows\System32\calc.exe
```

也可以不指定可执行文件的路径，系统约定从指定的路径找可执行文件，这个路径通过 PATH 环境变量指定。环境变量是一个"变量名 = 变量值"的对应关系，每一个变量都有一个或者多个值与之对应。如果是多个值，则这些值之间用分号分开，例如 PATH 环境变量可能对应这样的值："C:\Windows\system32;C:\Windows"，表示 Windows 会从 C:\Windows\system32 和 C:\Windows 两个路径找可执行文件。

设置或者修改环境变量的具体操作步骤是：首先在 Windows 桌面右击此电脑→属性→高级系统设置→环境变量，然后设置用户变量，或者系统变量，然后再设置环境变量 PATH 的值。

其实打开桌面上我的电脑就是运行资源管理器。打开资源管理器的另外一种方法是：首先按住键盘上的窗口键不放，然后再按 E 键。

需要重新启动命令行才能让环境变量设置生效。为了检查环境变量是否设置正确，可以在命令行中显示指定环境变量的值，例如显示 PATH 的值：

```
C:\Users\Administrator>PATH
```

为了从命令行修改环境变量，使用 Chocolatey 安装环境变量编辑器 Rapid Environment Editor。

```
>choco install rapidee
```

服务器的名称通过 DNS 服务器转换成对应的 IP 地址，也就是说，通过 DNS 取得该 URL 域名的 IP 地址。电脑需要选择一个好的 DNS 服务器，常用的有 114.114.114.114，或者 8.8.8.8。

使用 WMIC（Windows Management Instrumentation Command-line）设置 DNS：

```
wmic nicconfig where (IPEnabled=TRUE) call SetDNSServerSearchOrder ("114.114. 114.114")
```

为了快速访问网站，还可以直接在本机设置网站的 IP 地址，例如，首先得到 stackoverflow.com 的 IP 地址：

```
C:\Users\Administrator>nslookup stackoverflow.com
```

服务器：public1.114dns.com

Address: 114.114.114.114

非权威应答：

名称：stackoverflow.com

Addresses: 151.101.1.69
          151.101.129.69
          151.101.193.69
          151.101.65.69

然后修改 HOSTS 文件，它其实就是一个文本文件。每行一个域名对应一个 IP 地址。

```
151.101.1.69        stackoverflow.com
```

1.4 上手 Scrapy 网络爬虫开发

Python 软件基金会维护的 Python 语言代码解释器，可以从 Python 官方网站 https://www.python.org 下载。

在 Windows 下安装 Python 以后，在控制台输入 python 命令进入交互式环境。

```
d:\data>python
Python 3.7.2 (tags/v3.7.2:9a3ffc0492, Dec 23 2018, 23:09:28) [MSC v.1916 64 bit (AMD64)] on win32
Type "help", "copyright", "credits" or "license" for more information.
>>>
```

由于开源软件的迅速发展，可以借助开源软件简化自然语言处理的开发工作。简单地，可以使用 Sublime 这样的文本编辑器写 Python 代码，也可以使用 Eric（https://eric-ide.python-projects.org）或者 Microsoft Visual Studio 这样的集成开发环境。

Scrapy（https://github.com/scrapy/scrapy）是一个流行的爬虫框架。要安装 Scrapy，请在终端使用以下命令：

```
pip install Scrapy
```

Scrapy shell 是一个交互式 shell，可以在其中非常快速地尝试和调试爬虫代码。通常，我们通过传递网页的 URL 来启动一个 shell，如下所示：

语法：scrapy shell <url_to_be_scraped>

例如：

scrapy shell http://quotes.toscrape.com/tag/friends/

获得网页标题：

```
In [1]: response.xpath('//title/text()').get()
Out[1]: 'Quotes to Scrape'
```

在浏览器中查看网页源代码，可以看到网页标题：

```html
<html lang="en">
<head>
  <meta charset="UTF-8">
  <title>Quotes to Scrape</title>
   <link rel="stylesheet" href="/static/bootstrap.min.css">
   <link rel="stylesheet" href="/static/main.css">
</head>
...
```

一旦我们学会了启动 shell，我们就可以用它来测试爬取代码。在编写任何 Python 爬虫代码之前，应该使用 shell 测试网页以进行抓取。Scrapy shell 有一些可用的快捷方式，一旦我们启动了 shell，它们就可用了。快捷方式介绍如下。

shelp()：shelp() 命令，显示 Scrapy 对象列表和有用的快捷方式。可以看到，Request 对象代表发送到链接 http://quotes.toscrape.com/tag/friends/ 的 GET 请求。此外，如果 Response 对象包含一个 200 HTTP 代码，表示请求成功，除此之外，它还提到了 Crawler 和 Spider 对象的位置。

fetch（URL）："URL"是指向需要抓取的网页的链接。fetch 快捷方式接受一个 URL，即要抓取的网页，它返回爬虫信息，以及响应是成功还是失败。在下面的示例中，我们有一个有效的 URL 和一个无效的 URL，根据请求的性质，fetch 会显示错误或成功代码。

```
In [4]: fetch('http://quotes.toscrape.com/tag/friends/')
2022-07-27 09:56:31 [scrapy.core.engine] DEBUG: Crawled (200) <GET http://quotes.toscrape.com/tag/friends/> (referer: None)

In [5]: fetch('http://quotes.toscrape.com/tafrnd/')
2022-07-27 09:57:10 [scrapy.core.engine] DEBUG: Crawled (404) <GET http://quotes.toscrape.com/tafrnd/> (referer: None)
```

fetch（request）：我们可以创建一个 Request 对象，并将其传递给 fetch() 方法，为此，需要创建一个 Scrapy 对象。Request 类提及到了所需的 HTTP 方法、网页的 URL、标头（如果有的话）。我们要抓取 URL='http://quotes.toscrape.com/tag/friends/'的网页，我们需要准备请求对象为：

```
fetch(request_object)
```

终端输入输出为：

```
In [6]: url = 'http://quotes.toscrape.com/tag/friends/'

In [7]: request = scrapy.Request(url,method='GET')

In [8]: fetch(request)
2022-07-27 11:07:40 [scrapy.downloadermiddlewares.retry] DEBUG: Retrying <GET http://quotes.toscrape.com/tag/friends/> (failed 1 times): [<twisted.python.failure.Failure twisted.internet.error.ConnectionLost: Connection to the other side was lost in a non-clean fashion.>]
2022-07-27 11:07:41 [scrapy.core.engine] DEBUG: Crawled (200) <GET http://quotes.toscrape.com/tag/friends/> (referer: None)
```

view（Response）：在默认浏览器中打开网页，网页是作为 Request 对象或 fetch() 方法中的 URL 发送的网页。当我们输入 view（Response）时，在上述 fetch（Request）之后，网页会在默认浏览器中打开。

1.5 本章小结

本章介绍了各种网络爬虫以及开发网络爬虫所需要的软件工具。最后，介绍了从 Scrapy shell 上手网络爬虫开发。

第 2 章 Python 开发快速入门

为了方便采用自底向上的方式实现网络爬虫，需要复习下 Python 编程基础，并准备好相关的算法基础，有了这些基础就可以实现简单的信息采集或者分布式网络爬虫系统。

需要熟练掌握 Python 开发方面的基础知识，才能使用 Python 实现网络爬虫。

2.1 变量

定义变量时不声明类型，但变量在内部是有类型的。在交互式环境下输入如下代码会输出变量 a 的类型：

```
>>> a='test'
>>> type(a)    #调用 type()函数得到变量 a 的类型
<class 'str'>
```

2.2 注释

和 shell 类似，Python 脚本中用 # 表示注释。但如果 # 位于第一行开头，并且是 #!（称为 Shebang）则例外，它表示该脚本使用后面指定的解释器 /usr/bin/python3 解释执行。每个脚本程序只能在开头包含这个语句。

为了能够在源代码中添加中文注释，需要把源代码保存成 UTF-8 格式。例如：

```
# -*- coding: utf-8 -*

from bs4 import BeautifulSoup
import requests
url = "https://www.tutorialspoint.com/index.htm"
req = requests.get(url)
soup = BeautifulSoup(req.text, "html.parser")   #解析网页
print(soup.title)
```

2.3 简单数据类型

本节介绍包括数值、字符串和数组在内的简单数据类型。

2.3.1 数值

Python 中有三种不同的数值类型：int（整数）、float（浮点数）和 complex（复数）。和 Java 或者 C 语言中的 int 类型不同，Python 中的 int 类型是无限精度的，例如：

```
>>> i=3243244444444444444444444444444444444444448797687567567657000000000000000000000000000000000000000000000000000000000000000564564
>>> i
3243244444444444444444444444444444444444448797687567567657000000000000000000000000000000000000000000000000000000000000000564564
>>> type(i)
<class 'int'>
```

Python 依据 IEEE 754 标准使用二进制表示 float（浮点数），所以存在表示精度的问题，例如：

```
>>> 0.1 == 0.10000000000000000000000001
True
```

可以导入 decimal 模块并使用十进制表示完整的小数，例如：

```
>>> import decimal
>>> a = decimal.Decimal('0.1')
>>> b = decimal.Decimal('0.10000000000000000000000001')
>>> a == b
False
```

在傅里叶变换中会用到复数。复数在 Python 中是一个基本数据类型（complex），例如：

```
>>> complex(2,3)
(2+3j)
```

一个复数有一些内置的访问器：

```
>>> z = 2+3j
>>> z.real
2.0
>>> z.imag
3.0
>>> z.conjugate()
(2-3j)
```

几个内置函数都支持复数运算:

```
>>> abs(3 + 4j)
5.0
>>> pow(3 + 4j, 2)
(-7+24j)
```

标准模块 cmath 具有处理复数的更多功能:

```
>>> import cmath
>>> cmath.sin(2 + 3j)
(9.15449914691143-4.168906959966565j)
```

用于数值运算的算术运算符说明列表如表 2-1 所示。

表 2-1 算术运算符

语　　法	数学含义	运算符名字
a+b	a+b	加
a-b	a-b	减
a*b	a×b	乘法
a/b	a÷b	除法
a//b	$\lfloor a \div b \rfloor$	地板除
a%b	a mod b	模
-a	-a	取负数
abs(a)	$\lvert a \rvert$	绝对值
a**b	a^b	指数
math.sqrt(a)	$\sqrt{a}$	平方根

对于 "/" 运算,就算分子分母都是 int,返回的也将是浮点数,例如:

```
>>> print(1/3)
0.3333333333333333
```

Python 支持不同的数字类型相加,它使用数字类型强制转换的方式来解决数字类型不一致的问题,就是说它会将一个操作数转换为与另一个操作数相同的数据类型。

如果有一个操作数是复数,则另一个操作数被转换为复数:

```
>>> 3.0 + (5+6j)    # 非复数转复数
(8+6j)
```

整数转换为浮点数:

```
>>> 6 + 7.0 # 非浮点型转浮点型
```

```
13.0
```

Python 代码中一般一行就是一条语句,但是可以使用斜杠(\)将一条语句分为多行显示。例子代码如下:

```
>>> a = 1
>>> b = 2
>>> c = 3
>>> total = a + \
... b + \
... c
>>> total
6
```

2.3.2 字符串

在计算机编程中,字符串是一个字符序列,例如,"hello" 是一个包含字符序列 'h'、'e'、'l'、'l' 和 'o' 的字符串。

我们使用单引号或双引号来表示 Python 中的字符串,例如:

```
# create a string using double quotes
string1 = "Python programming"

# create a string using single quotes
string1 = 'Python programming'
```

可以通过三种方式访问字符串中的字符。

(1)索引。一种方法是将字符串视为列表并使用索引值,例如:

```
greet = 'hello'

# access 1st index element
print(greet[1]) # "e"
```

(2)负索引。与列表类似,Python 允许对其字符串进行负索引,例如:

```
greet = 'hello'

# access 4th last element
print(greet[-4]) # "e"
```

(3)切片。使用切片运算符冒号(:)访问字符串中的字符范围,例如:

```
greet = 'Hello'
```

```
# access character from 1st index to 3rd index
print(greet[1:4])   # "ell"
```

可以在 Python 中创建多行字符串，为此，我们使用三个双引号 """ 或三个单引号 '''，例如：

```
# multiline string
message = """
Never gonna give you up
Never gonna let you down
"""

print(message)
```

在上面的示例中，封闭三引号内的任何内容都是一个多行字符串。

可以使用 strip() 方法去掉字符串首尾的空格或者指定的字符。

```
term = "   hi    ";
print(term.strip());   # 去除首尾空格
```

使用 split() 方法将句子分成单词。下面的 Mary 是一个单一的字符串，尽管这是一个句子，但这些词语并没有表示成严谨的单位，为此，需要一种不同的数据类型：字符串列表，其中每个字符串对应一个单词。使用 split() 方法来把句子切分成单词：

```
>>> mary = 'Mary had a little lamb'
>>> mary.split()
['Mary', 'had', 'a', 'little', 'lamb']
```

split() 方法根据空格拆分 mary，返回的结果是 mary 中的单词列表，此列表包含 len() 函数演示的 5 个项目。对于 mary，len() 函数返回字符串中的字符数（包括空格）。

```
>>> mwords = mary.split()
>>> mwords
['Mary', 'had', 'a', 'little', 'lamb']
>>> len(mwords)                    # mwords 中的项目数
5
>>> len(mary)                      # 字符数
22
```

空白字符包括空格 ' '，换行符 '\n' 和制表符 '\t' 等。split() 方法可以分隔这些字符的任何组合序列：

```
>>> chom = ' colorless     green \n\tideas\n'
>>> print(chom)
 colorless     green
	ideas
```

```
>>> chom.split()
['colorless', 'green', 'ideas']
```

通过提供可选参数，split('x') 可用于在特定子字符串 'x' 上拆分字符串。如果没有指定 'x'，split() 只是在所有空格上分割，如下所示。

```
>>> mary = 'Mary had a little lamb'
>>> mary.split('a')                    # 根据 'a' 切分
['M', 'ry h', 'd ', ' little l', 'mb']
>>> hi = 'Hello mother,\nHello father.'
>>> print(hi)
Hello mother,
Hello father.
>>> hi.split()                          # 没有给出参数：在空格上分割
['Hello', 'mother,', 'Hello', 'father.']
>>> hi.split('\n')                      # 仅在 '\n' 上分割
['Hello mother,', 'Hello father.']
```

但是如果你想将一个字符串拆分成一个字符列表呢？在 Python 中，字符只是长度为 1 的字符串。list() 函数将字符串转换为单个字符的列表：

```
>>> list('hello world')
['h', 'e', 'l', 'l', 'o', ' ', 'w', 'o', 'r', 'l', 'd']
```

如果有一个单词列表，可以使用 join() 方法将它们重新组合成一个单独的字符串。在"分隔符"字符串 'x' 上调用 'x'.join(y) 会连接列表 y 中由 'x' 分隔的每个元素。下面，mwords 中的单词用空格连接回句子字符串：

```
>>> mwords
['Mary', 'had', 'a', 'little', 'lamb']
>>> ' '.join(mwords)
'Mary had a little lamb'
```

也可以在空字符串 '' 上调用该方法作为分隔符，效果是列表中的元素连接在一起，元素之间没有任何内容。下面，将一个字符列表放回到原始字符串中：

```
>>> hi = 'hello world'
>>> hichars = list(hi)
>>> hichars
['h', 'e', 'l', 'l', 'o', ' ', 'w', 'o', 'r', 'l', 'd']
>>> ''.join(hichars)
'hello world'
```

对一个字符串取子串的示例代码如下：

```
>>> x = "Hello World!"
```

```
>>> x[2:]
'llo World!'
>>> x[:2]
'He'
>>> x[:-2]
'Hello Worl'
>>> x[-2:]
'd!'
>>> x[2:-2]
'llo Worl'
```

使用 ord() 函数和 chr() 函数实现字符串和整数之间的互相转换：

```
>>> a = 'v'
>>> i = ord(a)
>>> chr(i)
'v'
```

字符串插值是将变量的值替换为字符串中的占位符的过程，例如：

```
# Python program to demonstrate
# string interpolation

n1 = 'Hello'
n2 = 'GeeksforGeeks'

# f tells Python to restore the value of two
# string variable name and program inside braces {}
print(f"{n1}! This is {n2}")
```

2.3.3 数组

创建一个数组，然后向这个数组中添加元素的代码如下：

```
>>> temp_list = []
>>> print(temp_list)
[]
>>> temp_list.append("one")
>>> temp_list.append("two")
>>> print(temp_list)
['one', 'two']
>>>
```

创建一个指定长度的数组：

```
>>> size = 10
```

```
>>> lst = [None] * size
>>> lst
[None, None, None, None, None, None, None, None, None, None]
```

2.4 字面值

Python 包括如下几种类型的字面值。

（1）数字：整数、浮点数、复数；

（2）字符串：以单引号、双引号或者三引号定义字符串；

（3）布尔值：True 和 False；

（4）空值：None。

有 4 种不同的字面值集合，分别是：列表字面值、元组字面值、字典字面值和集合字面值，示例代码如下：

```
fruits = ["apple", "mango", "orange"]         # 列表
numbers = (1, 2, 3)                           # 元组
alphabets = {'a':'apple', 'b':'ball', 'c':'cat'}   # 字典
vowels = {'a', 'e', 'i', 'o', 'u'}            # 集合

print(fruits)
print(numbers)
print(alphabets)
print(vowels)
```

2.5 控制流

完成一件事情要有流程控制，例如，洗衣的 3 个步骤：把脏衣服放进洗衣机→等洗衣机洗好衣服→晾衣服，这是顺序控制结构。

顺序执行的代码采用相同的缩进，叫作一个代码块。Python 没有像 Java 或者 C# 语言那样采用 {} 分隔代码块，而是采用代码缩进和冒号来区分代码之间的层次。

缩进的空白数量是可变的，但是所有代码块语句必须包含相同的缩进空白数量。NodePad++ 这样的文本编辑器支持选择多行代码后，按 Tab 键改变代码块的缩进格式。

控制流用来根据运行时的情况调整语句的执行顺序。流程控制语句可以分为条件语句和迭代语句。

2.5.1 if 语句

当路径不存在就创建它，可以使用条件语句实现。条件语句的一般形式如下：

```
if 条件:
    语句 1
    语句 2...
elif 条件:
    语句 1
    语句 2...
else:
    语句 1
    语句 2...
语句 x
```

例如，判断一个数是否是正数：

```
x = -32.2;
isPositive = (x > 0);
if isPositive:
    print(x, " 是正数 ");
else:
    print(x, " 不是正数 ");
```

这里的 if 复合语句，首行以关键字开始，以冒号（:）结束。

使用关系运算符和条件运算符作为判断依据。关系运算符返回一个布尔值。关系运算符完整的列表如表 2-2 所示。

表 2-2 关系运算符

运 算 符	用 法	返回 true，如果……
>	a > b	a 大于 b
>=	a >= b	a 大于或等于 b
<	a < b	a 小于 b
<=	a <= b	a 小于或等于 b
==	a == b	a 等于 b
!=	a != b	a 不等于 b

如果要针对多个值测试一个变量，则可以在 if 条件判断中使用一个集合：

```
x = "Wild things"
y = "throttle it back"
```

```
z = "in the beginning"
if "Wild" in {x, y, z}: print (True)
```

2.5.2 循环

使用复印机复印一个证件,可以设定复制的份数,例如,复制 3 份副本。在 Python 中,可以使用 for 循环或者 while 循环实现多次重复执行一个代码块。

for 循环可以遍历任何序列,例如,输出数组中的元素:

```
mylist = [1,2,3]
for item in mylist:
    print(item)
```

可以使用 range() 函数循环一组代码指定的次数。range() 函数返回一个数字序列,默认从 0 开始,默认以 1 递增,并以指定的数字结束,例如:

```
for num in range(1, 23):
    url = f"https://slickdeals.net/computer-deals/?page={num}"
    print(url)
```

每一次在执行循环代码块之前,根据循环条件决定是否继续执行循环代码块,当满足循环条件时,继续执行循环体中的代码。在循环条件之前写上关键词 while,这里的 while 就是"当"的意思,例如,当用户直接输入回车时退出循环:

```
import sys

while True:
    line = sys.stdin.readline().strip()
    if not line:
        break
    print(line)
```

2.6 列表

可以使用一个列表(List)存储任何类型的对象,例如:

```
list1 = ['physics', 'chemistry', 1997, 2000];
list2 = [1, 2, 3, 4, 5, 6, 7 ];
print("list1[0]: ", list1[0])
print("list2[1:5]: ", list2[1:5])
```

输出:

```
list1[0]:  physics
list2[1:5]:  [2, 3, 4, 5]
```

此外，列表甚至可以将另一个列表作为项目，这称为嵌套列表。

```
my_list = ["mouse", [8, 4, 6], ['a']]     # 嵌套列表
```

使用 range 函数生成列表：

```
>>> list(range(10))                       # 从 0 开始到 10，步长为 1
[0, 1, 2, 3, 4, 5, 6, 7, 8, 9]
>>> list(range(0, 30, 5))                 # 从 0 开始到 30，步长为 5
[0, 5, 10, 15, 20, 25]
```

可以使用赋值运算符（=）来更改一个项目或项目范围。

```
odd = [2, 4, 6, 8]                        # 错误的值

odd[0] = 1                                # 改变第一项

print(odd)                                # 输出：[1, 4, 6, 8]

odd[1:4] = [3, 5, 7]                      # 改变第 2 至第 4 项

print(odd)                                # 输出：[1, 3, 5, 7]
```

可以使用 append() 方法将一个项添加到列表中，或使用 extend() 方法添加多个项。

```
odd = [1, 3, 5]

odd.append(7)

print(odd)                                # 输出：[1, 3, 5, 7]

odd.extend([9, 11, 13])

print(odd)                                # 输出：[1, 3, 5, 7, 9, 11, 13]
```

可以使用 + 运算符来连接两个列表。* 运算符重复列表给定次数。

```
odd = [1, 3, 5]

print(odd + [9, 7, 5])                    # 输出：[1, 3, 5, 9, 7, 5]

print(["re"] * 3)                         # 输出：["re", "re", "re"]
```

此外，我们可以使用方法 insert() 在所需位置插入一个项目，或者通过将多个项目挤压到列表的空白切片中来插入多个项目。

```
odd = [1, 9]
odd.insert(1,3)

print(odd)                      # 输出: [1, 3, 9]

odd[2:2] = [5, 7]

print(odd)                      # 输出: [1, 3, 5, 7, 9]
```

可以使用关键字 del 从列表中删除一个或多个项目。

```
my_list = ['p','r','o','b','l','e','m']

del my_list[2]                  # 删除一个项目

print(my_list)                  # 输出: ['p', 'r', 'b', 'l', 'e', 'm']

del my_list[1:5]                # 删除多个项目

print(my_list)                  # 输出: ['p', 'm']
```

甚至可以完全删除列表。

```
del my_list                     # 删除整个列表

print(my_list)                  # 错误: 列表未定义
```

可以使用 remove() 方法删除给定的项目, 或使用 pop() 方法删除给定索引处的项目, 也可以使用 clear() 方法清空列表。

```
my_list = ['p','r','o','b','l','e','m']
my_list.remove('p')

print(my_list)                  # 输出: ['r', 'o', 'b', 'l', 'e', 'm']

print(my_list.pop(1))           # 输出: 'o'

print(my_list)                  # 输出: ['r', 'b', 'l', 'e', 'm']

print(my_list.pop())            # 输出: 'm'

print(my_list)                  # 输出: ['r', 'b', 'l', 'e']

my_list.clear()

print(my_list)                  # 输出: []
```

最后，我们还可以通过为一个元素片段分配一个空列表来删除列表中的项目。

```
>>> my_list = ['p','r','o','b','l','e','m']
>>> my_list[2:3] = []
>>> my_list
['p', 'r', 'b', 'l', 'e', 'm']
>>> my_list[2:5] = []
>>> my_list
['p', 'r', 'm']
```

for-in 语句可以轻松遍历列表中的项目：

```
for fruit in ['apple','banana','mango']:
    print("I like",fruit)
```

为了复制出一个新的列表，可以使用内置的 list.copy() 方法（从 Python 3.3 开始提供）。

```
>>> old_list = [1, 2, 3]
>>> new_list = old_list.copy()
```

使用 new_list = my_list，实际上没有两个列表，赋值仅复制对列表的引用，而不是实际列表，因此 new_list 和 my_list 在赋值后引用相同的列表。

通常，我们只想收集符合特定条件的项目。下面，我们有一个单词列表，我们只想从中提取包含 'wo' 的单词，为此，我们需要先创建一个新的空列表，然后遍历原始列表以查找要放入的项目：

```
>>> wood = 'How much wood would a woodchuck chuck if a woodchuck could chuck wood?'.split()
>>> wood
['How', 'much', 'wood', 'would', 'a', 'woodchuck', 'chuck', 'if', 'a', 'woodchuck', 'could', 'chuck', 'wood?']
>>> wolist = []                              # 创建一个空的列表
>>> for x in wood:
        if 'wo' in x:
            wolist.append(x)                 # 向列表增加项目
>>> wolist
['wood', 'would', 'woodchuck', 'woodchuck', 'wood?']
```

打印列表的内容：

```
>>> mylist = ['x', 3, 'b']
>>> print('[%s]' % ', '.join(map(str, mylist)))
[x, 3, b]
```

2.7 元组

元组是一个不可变的 Python 对象序列。元组变量的赋值要在定义时就进行，赋值之后就不允许有修改。

```
tup1 = ('physics', 'chemistry', 1997, 2000);
tup2 = (1, 2, 3, 4, 5, 6, 7 );
print( "tup1[0]: ", tup1[0]);
print( "tup2[1:5]: ", tup2[1:5]);
```

通常将元组用于异构（不同）数据类型，将列表用于同类（相似）数据类型。

包含多个项目的文字元组可以分配给单个对象。当发生这种情况时，就好像元组中的项目已经"打包"到对象中。

```
>>> t = ('foo', 'bar', 'baz', 'qux')
```

将元组中的元素分别赋给变量称为拆包。

```
>>> (s1, s2, s3, s4) = ('foo', 'bar', 'baz', 'qux')
>>> s1
'foo'
>>> s2
'bar'
>>> s3
'baz'
>>> s4
'qux'
```

包装和拆包可以合并为一个语句，以进行复合分配：

```
>>> (s1, s2, s3, s4) = ('foo', 'bar', 'baz', 'qux')
>>> s1
'foo'
>>> s2
'bar'
>>> s3
'baz'
>>> s4
'qux'
```

可以构建一个元组组成的数组：

```
>>> pairs = [("a", 1), ("b", 2), ("c", 3)]
>>> for a, b in pairs:
...     print(a, b)
```

```
...
a 1
b 2
c 3
```

可以使用命名元组给元组中的元素起一个有意义的名字:

```
import collections

# 声明一个名为 Person 的命名元组，这个元组包含 name 和 age 两个键
Person = collections.namedtuple('Person', 'name age')

# 使用命名元组
bob = Person(name='Bob', age=30)
print('\nRepresentation:', bob)

jane = Person(name='Jane', age=29)
print('\nField by name:', jane.name)

print('\nFields by index:')
for p in [bob, jane]:
    print('{} is {} years old'.format(*p))
```

2.8 集合

可以使用运算符 in 来检查给定元素是否存在于集合中。如果集合中存在指定元素，则返回 True，否则返回 False。

```
>>> s = {1,2,3,4,5}              # 创建集合对象并将其分配给变量 s
>>> contains = 1 in s            # 判断是否包含的例子
>>> print(contains)
True
>>> contains = 6 in s
>>> print(contains)
False
```

输出字符串 'banana' 中的字符集合:

```
>>> set(c for (i,c) in enumerate('banana'))
{'n', 'a', 'b'}
```

2.9 字典

字典是另一种可变容器模型,且可存储任意类型的对象。要访问字典元素,可以使用熟悉的方括号和键来获取它的值。

```
dict = {'Name': 'Zara', 'Age': 7, 'Class': 'First'}
print("dict['Name']: ", dict['Name'])
print("dict['Age']: ", dict['Age'])
```

如果需要根据字典中的值排序,由于字典本质上是无序的,所以可以把排序结果保存到有序的列表。

```
>>> x = {1: 2, 3: 4, 4: 3, 2: 1, 0: 0}
>>> sorted_by_value = sorted(x.items(), key=lambda kv: kv[1])
>>> print(sorted_by_value)
[(0, 0), (2, 1), (1, 2), (4, 3), (3, 4)]
```

OrderedDict 是一个字典子类,它会记住键/值对的顺序。

```
import collections

print('普通的字典:')
d = {}
d['a'] = 'A'
d['b'] = 'B'
d['c'] = 'C'

for k, v in d.items():
    print(k, v)

print('\n有序的字典:')
d = collections.OrderedDict()
d['a'] = 'A'
d['b'] = 'B'
d['c'] = 'C'
d['a'] = 'a'

for k, v in d.items():
    print(k, v)
```

2.10 函数

把一段多次重复出现的函数命名成一个有意义的名字,然后通过名字来执行这段代码。

有名字的代码段就是一个函数。使用关键字 def 定义一个函数,例如:

```
def square(number):              # 定义一个名为 square 的函数
    return number * number       # 返回一个数的平方
print(square(3))                 # 输出: 9
```

代码中可以给函数增加说明:

```
def square_root(n):
    """ 计算一个数字的平方根。

    Args:
        n: 用来求平方根的数字。
    Returns:
        n 的平方根。
    Raises:
        TypeError: 如果 n 不是数字。
        ValueError: 如果 n 是负数。

    """
    pass
```

参数可以有默认值,例如,定义一个名为 greet_person 的函数:

```
def greet_person(person, number=2):
    for greeting in range(number):
        print(f"Hello {person}! How are you doing today?")
# 1.
greet_person("Sara", 5)
# 2.
greet_person("Kevin")
```

输出结果如下:

```
Hello Sara! How are you doing today?
Hello Sara! How are you doing today?
Hello Sara! How are you doing today?
Hello Sara! How are you doing today?
Hello Sara! How are you doing today?
Hello Kevin! How are you doing today?
Hello Kevin! How are you doing today?
```

如果需要声明可变数量的参数,则在这个参数前面加 *,示例代码如下:

```
def myFun(*argv):
    for arg in argv:
        print (arg)
```

```
myFun('Hello', 'a', 'to', 'b')
```

函数定义中的特殊语法 **kwargs 用于传递一个键/值对的可变长度的参数列表，示例代码如下：

```
def myFun(**kwargs):
    for key, value in kwargs.items():
        print ("%s == %s" %(key, value))

# 调用函数
myFun(first ='test', mid ='for', last='abc')
```

输出结果如下：

```
first == test
mid == for
last == abc
```

每个 Python 文件/脚本（模块）都有一些未明确声明的内部属性。其中一个属性是 __builtins__ 属性，它本身包含许多有用的属性和功能，我们可以在这里找到 __name__ 属性，根据模块的使用方式，它可以具有不同的值。

当把 Python 模块作为程序直接运行时（无论是从命令行还是双击它），__name__ 中包含的值都是文字字符串 "__main__"。

相比之下，当一个模块被导入到另一个模块中（或者在 Python REPL 被导入）时，__name__ 属性中的值是模块本身的名称（即隐式声明它的 Python 文件/脚本的名称）。

Python 脚本执行的方式是自上而下的，指令在解释器读取它们时执行。这可能是一个问题，如果你想要做的就是导入模块并利用它的一个或两个方法。你会怎么做？你有条件地执行这些指令：将它们包装在一个 if 语句块中。

这是 'main 函数' 的目的，它是一个条件块，因此除非满足给定的条件，否则不会处理 main 函数。

main 函数的示例代码如下：

```
def main():
    print("Hello World!")


if __name__ == "__main__":
    main()
```

在 Python 中，函数是一级对象。这意味着函数就像其他任何对象一样，可以从函数返回函数。

在下面的程序中，我们定义了两个函数：function1() 和 function2()。function1() 返回 function2() 作为返回值。

```
def function1():
    return function2

def function2():
    print('Function 2')

x = function1()
x()
```

2.11 模块

可以使用 import 语句导入一个 .py 文件中定义的函数。一个 .py 文件就称之为一个模块（Module），例如存在一个 re.py 文件。可以使用 import re 语句导入这个正则表达式模块。使用正则表达式模块去掉一些标点符号的示例代码如下：

```
import re

line = 'Hi.'
normtext = re.sub(r'[\.,:;\?]', '', line)
print(normtext)
```

从 re 模块直接导入 sub 函数的示例代码：

```
from re import sub

line = 'Hi.'
normtext = sub(r'[\.,:;\?]', '', line)
print(normtext)
```

模块越来越多以后，会难以管理，例如，可能会出现重名的模块。一个班里有两个叫作陈晨的同学，如果他们在不同的小组，可以叫第一组的陈晨或者第三组的陈晨，这样就能区分同名了。为了避免名字冲突，模块可以位于不同的命名空间，叫作包。可以在模块名前面加上包名限定，这样即使模块名相同，也不会冲突了。

Python 中的外部模块也可以使用包管理器 pip 下载和安装，例如，安装模块 bs4：

```
pip install bs4
```

另一方面，一些模块，例如 Math 模块，不需要安装，我们只需要在 import 后加模块名称。

为了查看本地有哪些模块可用，可以在 Python 交互式环境中输入：

```
help('modules')
```

2.12 检查字符串是否包含子字符串

检查 Python 字符串是否包含子字符串的最简单方法是使用 in 运算符。

in 运算符用于在 Python 中检查数据结构的成员身份，它返回布尔值（True 或 False）。要使用 in 运算符检查字符串是否包含 Python 中的子字符串，我们只需在超字符串上调用它：

```
fullstring = "StackAbuse"
substring = "tack"

if substring in fullstring:
    print("Found!")
else:
    print("Not found!")
```

此运算符是调用对象的 __contains__ 方法的简写，也适用于检查列表中是否存在项。值得注意的是，它不是空安全的，因此如果 fullstring 指向 None，则会抛出异常：

```
TypeError: argument of type 'NoneType' is not iterable
```

为了避免这种情况，首先要检查它是否指向 None：

```
fullstring = None
substring = "tack"

if fullstring != None and substring in fullstring:
    print("Found!")
else:
    print("Not found!")
```

Python 中的 String 类型有一个名为 index() 的方法，可用于查找字符串中子字符串第一个出现的位置。

如果未找到子字符串，将引发 ValueError 异常，可以使用 try-except else 块来处理：

```
fullstring = "StackAbuse"
substring = "tack"

try:
    fullstring.index(substring)
```

```
except ValueError:
    print("Not found!")
else:
    print("Found!")
```

如果需要知道子字符串的位置，而不是只知道它在整个字符串中是否出现，则此方法非常有用。

String 类型有另一个方法 find()，它比 index() 更方便使用，因为我们不需要担心处理任何异常。

如果 find() 找不到匹配项，则返回 -1，否则返回较大字符串中子字符串的最左边索引。

```
fullstring = "StackAbuse"
substring = "tack"

if fullstring.find(substring) != -1:
    print("Found!")
else:
    print("Not found!")
```

如果希望避免捕获错误，那么该方法应该优于 index()。

正则表达式提供了一种更灵活（尽管更复杂）的方式来检查字符串的模式匹配。Python 附带了一个用于正则表达式的内置模块，称为 re。re 模块包含一个名为 search() 的函数，我们可以使用它来匹配子字符串模式：

```
from re import search

fullstring = "StackAbuse"
substring = "tack"

if search(substring, fullstring):
    print("Found!")
else:
    print("Not found!")
```

如果需要更复杂的匹配函数，如不区分大小写的匹配，则此方法是最好的，否则，对于简单的子字符串匹配用例，应该避免正则表达式的复杂性和较慢的速度。

2.13 面向对象编程

为了能够封装细节，需要抽象出对象。对象只是数据（变量）和作用于这些数据的方法（函数）的集合。类本质上是用于创建对象的模板。

就好像函数定义以关键字 def 开头一样，在 Python 中，可以使用关键字 class 定义一个类。

这是一个简单的类定义。

```
class MyNewClass:
    '''This is a docstring. I have created a new class'''
    pass
```

一个类创建一个新的本地命名空间，其中定义了所有属性。属性可以是数据或函数。其中还有一些特殊属性，这些属性以双下画线（__）开头，例如，__doc__ 为我们提供了该类的文档字符串，示例代码如下：

```
class MyClass:
    "This is my second class"
    a = 10
    def func(self):
        print('Hello')

print(MyClass.a)           # 输出: 10
print(MyClass.func)        # 输出: <function MyClass.func at 0x0000000003079BF8>
print(MyClass.__doc__)     # 输出: This is my second class
```

可以根据类模板来创建对象，创建对象的过程类似于函数调用。

```
ob = MyClass()
```

这将创建一个名为 ob 的新实例对象。我们可以使用对象名称前缀来访问对象的属性。

```
ob.func()   # 输出: Hello
```

您可能已经注意到了类中函数定义的 self 参数，但是我们将该方法简单地称为 ob.func() 而没有任何参数，它仍然奏效。这是因为，只要对象调用其方法，对象本身就作为第一个参数传递。因此，ob.func() 转换为 MyClass.func(ob)。

方法与对象实例或类相关联，函数则不是。当 Python 调度（调用）一个方法时，它会将该调用的第一个参数绑定到相应的对象引用（对于大多数方法，这个参数通常称为 self）。

在 Python 中，除了用户定义的属性外，每个对象都有一些默认的属性和方法。要查看对象的所有属性和方法，可以使用内置的 dir() 函数。执行以下脚本可以查看 ob 对象的所有属性：

```
ob = MyClass()

print(dir(ob))
```

在 Python 中，静态变量是在类的所有实例之间共享的变量，而不是每个实例唯一的变量，它有时也被称为类变量，因为它属于类本身而不是类的任何特定实例。

静态变量通常用一个值初始化，就像实例变量一样，但是可以通过类本身而不是通过实例来访问和修改它们，示例代码如下：

```
class Example:
    staticVariable = 5


print(Example.staticVariable)       # Prints '5'

instance = Example()
print(instance.staticVariable)      # Prints '5'

instance.staticVariable = 6
print(instance.staticVariable)      # Prints '6'
print(Example.staticVariable)       # Prints '5'

Example.staticVariable = 7
print(Example.staticVariable)       # Prints '7'
```

2.14　泛型

泛型类型允许我们定义可以处理不同数据类型的类、函数或方法，而无须事先指定确切的数据类型。当我们想要创建一个可以处理多种数据类型的单个实现时，这很有用。

要在 Python 中定义泛型类型，我们可以使用内置的模块 typing，它为 Python 3.5 及更新版本提供了一组类型提示。typing 模块定义了许多泛型类型，如 List、Tuple、Dict 和 Union，这些类型可用于定义泛型函数、方法和类。

在生活中，一般把同一个容器存放同一类东西，例如，棉签盒专门用来放棉签，糖果盒专门用来放糖果。可以使用泛型来检查集合中存储的数据类型，例如，List<str> 指定放字符串。

在这个基本示例中，我们将定义一个可以处理不同类型列表的泛型函数：

```
from typing import List, TypeVar


T = TypeVar('T')


def reverse_list(lst: List[T]) -> List[T]:
    return lst[::-1]
```

```
# Example usage
num_lst = [1, 2, 3, 4, 5]
str_lst = ['a', 'b', 'c', 'd', 'e']
print(reverse_list(num_lst))
print(reverse_list(str_lst))
```

定义一个可以使用不同类型值的泛型类:

```
from typing import TypeVar

T = TypeVar('T')

class Box:
    def __init__(self, value: T):
        self.value = value

    def get_value(self) -> T:
        return self.value

# Example usage
box1 = Box(10)
box2 = Box('Sling Academy')
print(box1.get_value())
print(box2.get_value())
```

创建一个通用的数据存储库类,该类可以处理不同的数据类型。

```
from typing import TypeVar, Generic, List

T = TypeVar('T')

class DataRepository(Generic[T]):
    def __init__(self):
        self.data = []

    def add_data(self, item: T) -> None:
        self.data.append(item)

    def remove_data(self, item: T) -> None:
        self.data.remove(item)

    def get_all_data(self) -> List[T]:
        return self.data

# Example usage
```

```
repo = DataRepository[int]()
repo.add_data(10)
repo.add_data(20)
repo.add_data(30)
print(repo.get_all_data()) # Output: [10, 20, 30]
repo.remove_data(20)
print(repo.get_all_data()) # Output: [10, 30]

repo2 = DataRepository[str]()
repo2.add_data('apple')
repo2.add_data('banana')
repo2.add_data('orange')
print(repo2.get_all_data()) # Output: ['apple', 'banana', 'orange']
repo2.remove_data('banana')
print(repo2.get_all_data()) # Output: ['apple', 'orange']
```

DataRepository 类使用泛型类型 T 来定义存储库可以存储的数据类型。泛型类型 T 确保添加到存储库的数据是正确的类型，泛型返回类型 List[T] 确保从存储库检索的数据也是正确的类型。

2.15　日志记录

网络爬虫的运行过程可能很长，为了监控运行状态，可以用日志记录。

```
import logging

logging.basicConfig(level=logging.DEBUG)
logging.debug('crawling...')
```

日志级别大小关系为：CRITICAL > ERROR > WARNING > INFO > DEBUG > NOTSET，当然也可以自己定义日志级别。

处理器将日志记录发送到任何输出，这些输出用自己的方式处理日志记录。

例如，FileHandler 会获取日志记录并将其附加到文件中。

标准日志记录模块已经配备了多个内置处理器，如：

（1）可以写入文件的多个文件处理器（TimeRotated、SizeRotated、Watched）；

（2）StreamHandler 可以输出到 stdout 或 stderr 等流；

（3）SMTPHandler 通过电子邮件发送日志记录；

（4）SocketHandler 将日志记录发送到流套接字。

此外，还有 SyslogHandler、NTEventHandler、HTTPHandler、MemoryHandler 等处理器。

格式器负责将元数据丰富的日志记录序列化为一个字符串。如果没有提供，则有一个默认格式器。记录库提供的通用格式器类将模板和样式作为输入，然后可以为日志记录对象中的所有属性声明占位符。

举个例子，'%(asctime)s %(levelname)s %(name)s: %(message)s' 会生成类似这样的日志：2017-07-19 15:31:13,942 INFO parent.child: Hello EuroPython。

请注意，属性消息是使用提供的参数对日志的原始模板进行插值的结果，例如，对于 logger.info("Hello %s", "Laszlo")，消息将是 "Hello Laszlo"。

TestStreamHandler.py 中的示例代码如下：

```
import logging

logger = logging.getLogger(__name__)
logger.setLevel(logging.INFO)
handler = logging.StreamHandler()
handler.setLevel(logging.INFO)
formatter = logging.Formatter('%(asctime)s [%(filename)s:%(lineno)s - %(funcName)s - %(levelname)s ] %(message)s')
handler.setFormatter(formatter)
logger.addHandler(handler)

string = ''
logger.info("trainning... \n {0}".format(string))
```

输出：

```
2022-06-24 22:01:28,465 [TestStreamHandler.py:12 - <module> - INFO ] trainning...
```

2.16　数据库

Python 编程语言具有强大的数据库编程功能。Python 支持各种数据库，如 SQLite、MySQL、Oracle、Sybase、PostgreSQL 等。数据库接口的 Python 标准是 Python DB-API，大多数 Python 数据库接口都遵循这个标准。

Python DB-API 包括的方法如下。

（1）module.connect()：连接到数据库，要连接的参数因模块而异。connect() 方法返回 Connection 对象或引发异常。

（2）Connection.cursor()：从连接生成游标对象。游标用于将 SQL 语句发送到数据库并获取结果。

（3）Connection.commit()：提交当前连接中所做的更改。如果要保存更改（如插入、更新或删除），必须在关闭连接之前调用 commit()。未提交的更改从当前连接中可见，但从其他连接中看不到。

（4）Connection.rollback()：回滚（撤销）当前连接中所做的更改。如果遇到异常时要继续使用同一连接，则必须回滚。

（5）Connection.close()：关闭连接。当程序退出时，连接总是隐式关闭的，但手动关闭连接是个好主意，尤其是当代码可能在循环中运行时。

（6）Cursor.execute(statement)：对数据库执行 SQL 语句。

（7）Cursor.execute(statement, tuple)：对数据库执行 SQL 语句。如果要将变量替换到 SQL 语句中，请使用这种形式。

（8）Cursor.fetchall()：从当前语句获取所有结果。

（9）Cursor.fetchone()：只获取一个结果，返回元组，如果没有结果，则返回 None。

SQLAlchemy 是一个开放源码的 SQL 工具包和对象关系映射器。

为了与数据库交互，我们需要获得它的句柄。session 对象是数据库的句柄。session 类是使用 sessionmaker() 定义的，这是一个可配置的 session 工厂方法，它绑定到前面创建的引擎对象，示例代码如下：

```
from sqlalchemy.orm import sessionmaker
import sqlalchemy as sa
from sqlalchemy import Column, Integer, String
from sqlalchemy.ext.declarative import declarative_base

#For learning purposes, we normally use a SQLite memory-only database for convenience.
engine = sa.create_engine("sqlite:///:memory:")
Session = sessionmaker(bind = engine)
session = Session()

Base = declarative_base()

class Customers(Base):
    __tablename__ = 'customers'

    id = Column(Integer, primary_key=True)
    name = Column(String)
    address = Column(String)
    email = Column(String)

Base.metadata.create_all(engine)

c1 = Customers(name = 'Ravi Kumar', address = 'Station Road Nanded', email = 'ravi@gmail.com')
```

```
session.add(c1)
session.commit()
```

2.17 本章小结

本章介绍了使用 Python 开发网络爬虫所需要的 Python 基础。

Python 于 20 世纪 80 年代后期由荷兰的 Guido van Rossum 设计，作为 ABC 语言的继承者，能够处理异常并与阿米巴操作系统连接。Python 2.0 于 2000 年 10 月 16 日发布，具有许多主要的新功能，包括循环检测垃圾收集器和对 Unicode 的支持。Python 3.0 于 2008 年 12 月 3 日发布，它是该语言的一个重要修订，并非完全向后兼容，它的许多主要功能都被反向移植到 Python 2.6.x 和 Python 2.7.x 版本系列。Python 3 的发布包括 2to3 实用程序，它可以自动（至少部分地）将 Python 2 的代码转换为 Python 3。

Python 是一种多范式编程语言。Python 完全支持面向对象的编程和结构化编程，其许多功能支持函数编程和面向切面编程。

Monty Python 引用经常出现在 Python 的代码和文化中，例如，Python 中经常使用的伪变量是 spam 和 eggs，而不是传统的 foo 和 bar。

第 3 章　使用 Python 开发网络爬虫

采购人员需要找到提供产品的有竞争力的厂家和价格，金融交易人员需要找到有潜力的投资公司，出版行业人士需要找到迅速变热的话题，这些都可以使用网络爬虫帮忙实现。

网络爬虫从互联网源源不断地抓取海量信息，搜索引擎结果中的信息都来源于此。如果把互联网比喻成一个覆盖地球的蜘蛛网，那么抓取程序就是在网上爬来爬去的蜘蛛。

网络爬虫爬取到的是网页的源文件，即 HTML 代码，抓到之后是从这个源码里面提取信息。

需要抓取哪些信息呢？应当首先关注一些高质量的网页信息。哪些是高质量的网页呢？网民投票选择出来的网页，也就是访问量高的网站中的一些热门网页。但是尺有所短，寸有所长，很多访问量一般的网站包括了更多问题的答案，有点类似长尾效益。Alexa（http://www.alexa.com）专门统计网站访问量并发布网站世界排名。

有些文档的时效性很强，例如新闻或者财经信息。大部分人想要知道的是当天股票市场的报道，只有很少人关心昨天的市场发生了什么。

网络爬虫需要实现的基本功能包括下载网页以及对 URL 地址的遍历。为了高效快速遍历网站还需要应用专门的数据结构来优化。爬虫很消耗带宽资源，设计爬虫时需要仔细考虑如何节省网络带宽。

3.1　使用 BeautifulSoup 实现定向采集

安装所需的第三方库：

```
pip install requests
pip install html5lib
pip install bs4
```

从网址访问 HTML 内容：

```
import requests
URL = "https://www.geeksforgeeks.org/data-structures/"
r = requests.get(URL)
print(r.content)
```

让我们试着理解这段代码。

（1）首先导入 requests 库。

（2）然后，指定要抓取的网页的 URL。

（3）向指定的 URL 发送 HTTP 请求，并将来自服务器的响应保存在名为 r 的响应对象中。

（4）现在，通过 print（r.content）来获取网页的原始 HTML 内容，它是字符串类型。

有时可能会收到"不接受"错误，这时请尝试添加如下浏览器用户代理。

```
headers = {'User-Agent': "Mozilla/5.0 (Windows NT 10.0; Win64; x64) AppleWebKit/537.36 (KHTML, like Gecko) Chrome/42.0.2311.135 Safari/537.36 Edge/12.246"}
# Here the user agent is for Edge browser on windows 10. You can find your browser
# user agent from the above given link.
r = requests.get(url=URL, headers=headers)
print(r.content)
```

解析 HTML 内容：

```
#This will not run on online IDE
import requests
from bs4 import BeautifulSoup

URL = "http://www.values.com/inspirational-quotes"
r = requests.get(URL)

soup = BeautifulSoup(r.content, 'html5lib') # If this line causes an error, run
# 'pip install html5lib' or install html5lib
print(soup.prettify())
```

BeautifulSoup 库的一个非常好的地方是它构建在 HTML 解析库（如 html5lib、lxml、html.parser 等）之上，因此可以同时创建 BeautifulSoup 对象和指定解析器库。在上面的例子中，

```
soup = BeautifulSoup(r.content, 'html5lib')
```

我们通过传递两个参数来创建一个 BeautifulSoup 对象。

（1）r.content：它是原始 HTML 内容。

（2）html5lib：指定我们要使用的 HTML 解析器。

现在 soup.prettify() 被打印出来了，它给出了从原始 HTML 内容创建的解析树的可视化表示。

现在，我们想从 HTML 内容中提取一些有用的数据。soup 对象包含嵌套结构中的所

有数据,这些数据可以通过编程方式提取。在我们的示例中,正在抓取一个包含一些名言的网页,因此,我们想创建一个程序来保存这些名言(以及有关它们的所有相关信息)。

```python
#Python program to scrape website
#and save quotes from website
import requests
from bs4 import BeautifulSoup
import csv

URL = "http://www.values.com/inspirational-quotes"
r = requests.get(URL)

soup = BeautifulSoup(r.content, 'html5lib')

quotes=[]  # a list to store quotes

table = soup.find('div', attrs = {'id':'all_quotes'})

for row in table.findAll('div',
                        attrs = {'class':'col-6 col-lg-4 text-center margin-30px-bottom sm-margin-30px-top'}):
    quote = {}
    quote['theme'] = row.h5.text
    quote['url'] = row.a['href']
    quote['img'] = row.img['src']
    quote['lines'] = row.img['alt'].split(" #")[0]
    quote['author'] = row.img['alt'].split(" #")[1]
    quotes.append(quote)

filename = 'inspirational_quotes.csv'
with open(filename, 'w', newline='') as f:
    w = csv.DictWriter(f,['theme','url','img','lines','author'])
    w.writeheader()
    for quote in quotes:
        w.writerow(quote)
```

在继续之前,建议浏览我们使用 soup.prettify() 方法打印的网页的 HTML 内容,并尝试找到导航到名言的模式或方法。

注意到所有的名言都在一个 id 为 'all_quotes' 的 div 容器中,因此,我们使用 find() 方法找到该 div 元素(在上面的代码中称为 table):

```
table = soup.find('div', attrs = {'id':'all_quotes'})
```

第一个参数是您要搜索的 HTML 标记;第二个参数是字典类型元素,用于指定与该标记关联的附加属性。find() 方法返回第一个匹配的元素。您可以尝试打印 table.prettify() 以了解这段代码的作用。

现在，在表格元素中，可以注意到每个名言都在一个类的 div 容器中，因此，我们遍历每个类为 quote 的 div 容器。在这里，我们使用 findAll() 方法，它在参数方面类似于 find() 方法，但它返回所有匹配元素的列表。现在，每个名言都使用名为 row 的变量进行迭代。

我们创建一个字典来保存有关名言的所有信息，可以使用点表示法访问嵌套结构。要访问 HTML 元素中的文本，我们使用 .text 文件：

```
quote['theme'] = row.h5.text
```

我们可以添加、删除、修改和访问标签的属性，这是通过将标签视为字典来完成的：

```
quote['url'] = row.a['href']
```

然后，所有名言都附加到名为 quotes 的列表中。

最后，我们希望将所有数据保存在某个 CSV 文件中。

```
filename = 'inspirational_quotes.csv'
with open(filename, 'w', newline='') as f:
    w = csv.DictWriter(f,['theme','url','img','lines','author'])
    w.writeheader()
    for quote in quotes:
        w.writerow(quote)
```

在这里，我们创建了一个 CSV 文件，名为 inspirational_quotes.csv，并将所有名言保存在其中以供进一步使用。

这就是一个如何在 Python 中创建网络爬虫的简单示例。从这里，您可以尝试抓取您选择的任何网站。

3.2 URL 基础

网络资源一般是 Web 服务器上的一些各种格式的文件，通过 URL 可以找到这些网络资源。URL 就是网络资源地址的缩写。

所谓网页抓取，就是把 URL 地址中指定的网络资源从网络流中读取出来，保存到本地，类似于使用程序模拟网页浏览器的功能。在 HTTP 请求中说明要哪个 URL，然后读取服务器端返回的资源内容。

3.2.1 URI

URI 包括 URL 和 URN，但是 URN 没有流行起来，只需要知道 URL 是 URI 的一种就可以。URL 是 Uniform Resource Locator 的缩写，译为"统一资源定位符"。通俗地说，URL 是 Internet 上描述信息资源的字符串，主要用在各种 WWW 客户程序和服务器程序上。采用 URL 可以用一种统一的格式来描述各种信息资源，包括文件、服务器的地址和目录等。

URL 由协议名、主机名和资源路径 3 部分组成，一个具体的示例如图 3-1 所示。

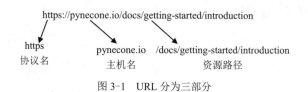

图 3-1　URL 分为三部分

协议也称为服务方式。

主机名也可以用主机 IP 地址代替，如果访问者能记住的话。有时也包括端口号，如果这个端口号访问者也能记住的话。

资源路径是主机资源的具体地址，如目录和文件名等。

协议名和主机名之间用"://"符号隔开，主机名和资源路径用"/"符号隔开。协议名和主机名是不可缺少的，资源路径如果正好是访问根路径下的缺省资源则有时可以省略。

例如：http://bj.cityhouse.cn/street/hd/onedistlist.html，其主机域名为 bj.cityhouse.cn。超级文本文件（文件类型为 .html）是在目录 street/hd 下的 onedistlist.html 中。

根据网址生成一个对应的 URI 对象：

```
>>> from urllib.parse import urlparse
>>> url = 'https://mail.google.com/mail/u/0/?tab = rm#inbox'
>>> t = urlparse(url)
ParseResult(scheme = 'https', netloc = 'mail.google.com', path = '/mail/u/0/', params = '', query = 'tab = rm', fragment = 'inbox')
```

3.2.2 解析相对地址

在 Windows 的控制台窗口中，可以根据当前路径的相对路径转移到一个路径，例如 cd .. 转移到当前路径的上级路径。在 HTML 网页中也经常使用相对 URL。

绝对 URL 就是不依赖其他的 URL 路径，例如："https://stackoverflow.com/questions/3764291"。在一定的上下文环境下可以使用相对 URL。网页中的 URL 地址可能是相对地址，例如："./index.html"。可以在 <A> 和 <img> 标签中使用相对 URL，例如：

```
<img src="../images/email.gif" />
```

可以根据所在网页的绝对 URL 地址，把相对地址转换为绝对地址。为了灵活地引用网站内部资源，相对路径在网页中很常见。爬虫为了后续处理方便，需要把相对地址转换为绝对地址。下面的代码把相对地址转换成绝对地址：

```
from requests.compat import urljoin
base='https://stackoverflow.com/questions/3764291'
print (urljoin(base,'..'))
```

3.2.3 DNS 解析

dig（domain information groper）是一个类 UNIX 网络管理命令行工具，用于查询域名系统（DNS）服务器。

```
# dig example.com
```

在任何 DNS 记录文件（Domain Name System (DNS) Zone file）中，都是以 SOA（Start of Authority）记录开始的。SOA 资源记录表明此 DNS 名称服务器是该 DNS 域中的数据的信息的最佳来源。

使用 dig 命令显示 SOA：

```
# dig SOA  example.com
```

DNS 主要使用用户数据报协议。用户数据报协议简称 UDP（User Datagram Protocol），使用端口 53 服务请求。DNS 查询由一个单一的来源于客户端的 UDP 请求和一个服务器返回的 UDP 答复组成。当响应数据超过 512 字节时，使用 TCP。有些解析器实现对所有的查询都使用 TCP。

在 Windows 下 DNS 解析的问题可以用 nslookup 命令来分析，例如：

```
C:\Users\Administrator>nslookup example.com
服务器:  pdns.dnspod.cn
Address:  119.29.29.29

非权威应答:
名称:    example.com
Addresses:  2606:2800:220:1:248:1893:25c8:1946
          93.184.216.34
```

如果想要在 Windows 下使用 dig 命令，则可以先安装 bind-toolsonly 包。

```
>choco install -y bind-toolsonly
```

根据服务器名称取得 IP 地址的代码如下。

```
# 先安装模块：
#pip install dnspython
import dns.resolver

result = dns.resolver.resolve('tutorialspoint.com', 'A')
for ipval in result:
    print('IP', ipval.to_text())
```

3.3 网络爬虫抓取原理

既然所有的网页都可能链接到其他的网站，那么从一个网站开始，跟踪所有网页上的所有链接，就可能遍历整个互联网。

为了更快地抓取想要的信息，网页抓取首先从一个已知的 URL 地址列表开始遍历，对垂直搜索来说，一般是积累的行业内的网站。有人可能会奇怪像 Google 或百度这样的搜索门户怎么设置这个初始的 URL 地址列表。一般来说，网站拥有者会把网站提交给分类目录，例如 dmoz（https://www.dmoz-odp.org/），爬虫则可以从开放式分类目录 dmoz 中抓取。

抓取下来的网页中包含了想要的信息，一般存放在数据库或索引库这样的专门的存储系统中，如图 3-2 所示。

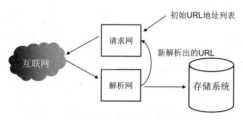

图 3-2 爬虫基本结构图

在搜索引擎中，爬虫程序是从一系列种子链接把这些初始的网页中的 URL 提取出来，放入 URL 工作队列（Todo 队列，又叫作 Frontier），然后遍历所有工作队列中的 URL，下载网页并把其中新发现的 URL 再次放入工作队列。为了判断一个 URL 是否已经遍历过，把所有遍历过的 URL 放入历史表（Visited 表）。爬虫抓取的基本过程如图 3-3 所示。

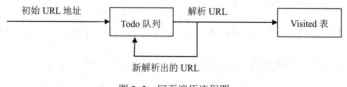

图 3-3 网页遍历流程图

Crawler 类的 Crawling 方法执行抓取过程，Crawler 类的主要实现代码如下。

```python
import logging
from urllib.parse import urljoin
import requests
from bs4 import BeautifulSoup
from collections import deque

logging.basicConfig(
    format='%(asctime)s %(levelname)s:%(message)s',
    level=logging.INFO)

class Crawler:

    def __init__(self, urls=deque()):
        self.visited_urls = set()
        self.urls_to_visit = urls

    def download_url(self, url):
        return requests.get(url).text

    def get_linked_urls(self, url, html):
        soup = BeautifulSoup(html, 'html.parser')
        for link in soup.find_all('a'):
            path = link.get('href')
            if path and path.startswith('/'):
                path = urljoin(url, path)
            yield path

    def add_url_to_visit(self, url):
        if url not in self.visited_urls and url not in self.urls_to_visit:
            self.urls_to_visit.append(url)

    def crawl(self, url):
        html = self.download_url(url)
        for url in self.get_linked_urls(url, html):
            self.add_url_to_visit(url)

    def run(self):
        while self.urls_to_visit:
            url = self.urls_to_visit.popleft()
            logging.info(f'Crawling: {url}')
            try:
                self.crawl(url)
            except Exception:
```

```
                logging.exception(f'Failed to crawl: {url}')
            finally:
                self.visited_urls.add(url)

if __name__ == '__main__':
    Crawler(urls=deque(['https://www.imdb.com/'])).run()
```

这里采用 Deque（双端队列）实现 Todo 队列。如果采用 Queue 来实现 Todo，则对每个增加到 Todo 的元素都需要用对象封装。Deque 允许在末端增加或删除元素，因为 Deque 底层采用数组实现，所以增加到 Deque 的元素不需要用对象封装。而且 Deque 性能比 Queue 更好。所以用 Deque 来实现 Todo 队列。

Visited 集合也叫作 URLSeen 存储。如果 Visited 是全局唯一的，那就需要同步了。

3.4 爬虫架构

本节首先介绍爬虫的基本架构，然后介绍可以在多台服务器上运行的分布式爬虫架构。

3.4.1 基本架构

一般的爬虫软件，通常都包含以下几个模块：
（1）保存种子 URL 和待抓取的 URL 的数据结构。
（2）保存已经抓取过的 URL 的数据结构，防止重复抓取。
（3）页面获取模块。
（4）对已经获取的页面内容的各个部分进行抽取的模块，例如抽取 HTML、JavaScript 等。
其他可选的模块包括：
（1）负责连接前处理模块。
（2）负责连接后处理模块。
（3）过滤器模块。
（4）负责多线程的模块。
（5）负责分布式的模块。
各模块详细介绍如下：

1. 保存种子和待抓取的 URL 的数据结构

农民会把有生长潜力的籽用作种子，这里把一些活跃的网页用作种子 URL，例如网站的首页或者列表页，因为在这些页面经常会发现新的链接。通常，爬虫都是从一系列的种子 URL 开始爬取，一般从数据库表或者配置文件中读取这些种子 URL。种子 URL 描述表如下。

表 3-1　种子 URL 描述表

字　段　名	字段类型	说　　明
Id	NUMBER(12)	唯一标识
url	Varchar(128)	网站 URL
source	Varchar(128)	网站来源描述
rank	NUMBER(12)	网站 PageRank 值

但是保存待抓取的 URL 的数据结构却因系统的规模、功能不同而可能采用不同的策略。一个比较小的爬虫程序，可能就使用内存中的一个队列，或者是优先级队列进行存储。一个中等规模的爬虫程序，可能使用 BekerlyDB 这种内存数据库来存储，如果内存中存放不下，还可以序列化到磁盘上。但是，真正的大规模爬虫系统，是通过服务器集群来存储已经爬取出来的 URL 的，并且，还会在存储 URL 的表中附带一些其他信息，比如说 PageRank 值等，供之后的计算用。

2. 保存已经抓取过的 URL 的数据结构

已经抓取过的 URL 的规模和待抓取的 URL 的规模是一个相当的量级，正如我们前面介绍的 Todo 队列和 Visited 表。但是，它们唯一的不同是，Visited 表会经常被查询，以便确定发现的 URL 是否已经处理过。因此，Visited 表数据结构如果是一个内存数据结构，可以采用散列表（HashMap 或者 HashSet）来存储；如果保存在数据库中，可以对 URL 列建立索引。

3. 页面获取模块

当从种子 URL 队列或者抓取出来的 URL 队列中获得 URL 后，便要根据这个 URL 来获得当前页面的内容，获得的方法非常简单，就是普通的 IO 操作。在这个模块中，仅仅是把 URL 所指的内容按照二进制的格式读出来，而不对内容做任何处理。

4. 提取已经获取的网页的内容中的有效信息

从页面获取模块的结果是一个表示 HTML 源代码的字符串。从这个字符串中抽取各种相关的内容，是爬虫软件的目的，因此，这个模块就显得非常重要。

通常，在一个网页中，除了包含有文本内容还有图片、超链接等。对于文本内容，首先把 HTML 源代码的字符串保存成 HTML 文件即可。关于超链接的提取，可以根据 HTML 语法，使用正则表达式来提取，并且把提取的超链接加入到 Todo 队列中，也可以使用专门的 HTML 文档解析工具。

在网页中，超链接不光指向 HTML 页面，还会指向各种文件，对于除了 HTML 页面的超链接之外，其他内容的链接不能放入 Todo 队列中，而要直接下载。因此，在这个模块中，还必须包含提取图片、JavaScript、PDF、DOC 等内容的部分。并且，在提取过程中，还要针对 HTTP 协议，来处理返回的状态码。这章我们主要研究网页的架构问题，将在下一章详细研究从各种文件格式提取有效信息。

5. 负责连接前处理模块，负责连接后处理模块，过滤器模块

如果只抓取某个网站的网页，则可以对 URL 按域名过滤。

6. 多线程模块

爬虫主要消耗三种资源：网络带宽、中央处理器和磁盘。三者中任何一者都有可能成为瓶颈，其中网络带宽一般是租用的，所以价格相对昂贵。为了增加爬虫效率，最直接的方法就是使用多线程的方式进行处理。在爬虫系统中，将要处理的 URL 队列往往是唯一的。多个线程顺序地从队列中取得 URL，之后各自进行处理（处理阶段是并发进行）。通常，可以利用线程池来管理线程。程序中最大可以使用的线程数是可配置的。

7. 分布式处理

分布式是当今计算的主流，这项技术也可以同时用在网络爬虫上面。后续有章节专门介绍多台机器并行采集的方法。

3.4.2　分布式爬虫架构

把抓取任务分布到不同的节点分布主要是为了可扩展性，也可以使用物理分布的爬虫系统，让每个爬虫节点抓取靠近它的网站，例如，北京的爬虫节点抓取北京的网站，上海的爬虫节点抓取上海的网站。还比如，电信网络中的爬虫节点抓取托管在电信的网站，联通网络中的爬虫节点抓取托管在联通的网站。

图 3-4 所示是一种没有中央服务器的分布式爬虫结构。

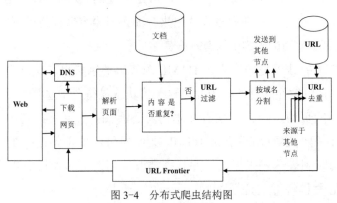

图 3-4　分布式爬虫结构图

要点在于按域名分配采集任务。每台机器扫描到的网址，不属于它自己的会交换给属于它的机器，例如，专门有一台机器抓取 s 开头的网站：http://www.sina.com.cn 和 http://www.sohu.com，而另外一台机器抓取 c 开头的网站：http://www.cctv.com。

垂直信息分布式抓取的基本设计如下。

（1）按要处理的信息的首字母做散列，让不同的机器抓取不同的信息。

（2）每台机器通过配置文件读取自己要处理的字母。每台机器抓取完一条结果后把该结果写入到统一的一个数据库中，比如说有 26 台机器，第一台机器抓取字母 a 开头的公司，第二台机器抓取字母 b 开头的公司，依次类推。

（3）如果某一台机器抓取速度太慢，则把该任务拆分到其他的机器。

3.4.3 垂直爬虫架构

垂直爬虫往往抓取指定网站的新闻或论坛等信息。可以指定初始抓取的首页或者列表页，然后提取相关的详细页中的有效信息存入数据库，总体结构如图 3-5 所示。

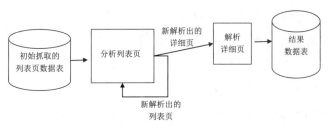

图 3-5 垂直爬虫结构图

垂直爬虫涉及的功能如下。

（1）从首页提取不同栏目的列表页。

（2）网页分类：把网页分类成列表页或详细页或者未知类型。

（3）列表页链接提取：从列表页提取同一个栏目下的列表页，这些页面往往用"下一页""尾页"等信息描述。

（4）详细页面内容提取：从详细页提取网页标题、主要内容、发布时间等信息。

每个网站可以用一个线程抓取，这样方便控制对被抓网站的访问频率。最好有通用的信息提取方式来解析网页，这样可以减少人工维护成本。同时，也可以采用专门的提取类来处理数据量大的网站，这样可以提高抓取效率。

3.5 下载网页

下载网页最基本的方法，可以用命令行工具软件 curl 下载网页。安装 curl 的一个方法是在命令行运行 choco install curl。

这里首先介绍网址的基本知识，然后介绍通过 curl 命令下载网页。

例如想要获得每个城市地区对应的街道，网址：

```
http://bj.cityhouse.cn/street/hd/onedistlist.html
```

其中包含一些这样的信息。先用网页浏览器打开这个网页，下载网页和浏览器根据网址打开网页的道理是一样的。"打开"网页的过程其实就是浏览器作为一个浏览的"客户端"向服务器端发送了一次请求，把服务器端的文件"抓"到本地，再进行解释、展现。更进一步，可以通过浏览器端查看"下载"过来的文件源代码。选择浏览器的"查看"|"源文件"菜单，就会看到浏览器从服务器上面"下载"下来的文件的源代码。

在上面的例子中，我们在浏览器的地址栏中输入的字符串叫作 URL。那么，什么是 URL 呢？直观地讲，URL 就是在浏览器端输入的 http://bj.cityhouse.cn/street/hd/onedistlist.html 这个字符串。用 URL 来代表一个网页。

首先要知道网页在哪个网站。每个网站都有一个英文名字，叫作域名，也就是主机名。然后要知道网页位于这个主机的哪个路径。还有，通过什么协议得到这个网页。所以 URL 由协议名、主机名和资源路径 3 部分组成。

一般是通过 HTTP 或者 HTTPS 协议得到网页内容，所以很多网址都用 http:// 或者 https:// 开头。

3.5.1 HTTP

一般通过 HTTP 协议和 Web 服务器打交道，这样的 Web 服务器又叫作 HTTP 服务器。HTTP 服务器存储了互联网上的数据并且根据 HTTP 客户端的请求提供数据。网络爬虫也是一种 HTTP 客户端，更常见的 HTTP 客户端是 Web 浏览器。客户端发起一个到服务器上指定端口（默认端口为 80）的 HTTP 请求，服务器端按指定格式返回网页或者其他网络资源，如图 3-6 所示。

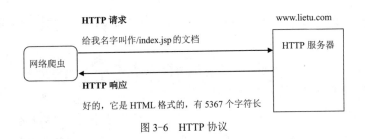

图 3-6 HTTP 协议

客户端向服务器发送的请求头包含请求的方法、URL、协议版本以及请求修饰符、客户信息和内容。服务器以一个状态行作为响应，相应的内容包括消息协议的版本、成功或者错误的编码加上服务器信息、实体元信息以及可能的实体内容。

HTTP 请求格式是：

```
<request line>
<headers>
<blank line>
[<request-body>]
```

在 HTTP 请求中，第一行必须是一个请求行（Request Line），用来说明请求类型、要访问的资源以及使用的 HTTP 版本。紧接着是头信息（Header），用来说明服务器要使用的附加信息。在头信息之后是一个空行，在此之后可以添加任意的其他数据，这些附加的数据称之为主体（Body）。

HTTP 规范定义了 8 种可能的请求方法。爬虫经常用到 GET、HEAD 和 POST 3 种请求方法，分别说明如下。

（1）GET：检索 URI 中标识资源的一个简单请求，例如爬虫发送请求 GET /index.html HTTP/1.1。

（2）HEAD：与 GET 方法相同，服务器只返回状态行和头标，并不返回请求文档，例如用 HEAD 方法请求检查网页更新时间。

（3）POST：服务器接受被写入客户端输出流中的数据的请求。可以用 POST 方法来提交表单参数。

例如，请求头：

```
Accept: text/plain, text/html
```

客户端说明了可以接收文本类型的信息，最好不要发送音频格式的数据。

```
Referer: http://www.w3.org/hypertext/DataSources/Overview.html
```

代表从这个网页开始直到正在请求的网页。

```
Accept-Charset: GB2312,utf-8;q=0.7
```

每个语言后包括一个 q-value，表示用户对这种语言的偏好估计，缺省值是 1.0，1.0 也是最大值。

```
Keep-alive: 115
Connection: keep-alive
```

Keep-alive 是指在同一个连接中发出和接收多次 HTTP 请求，单位是毫秒。

介绍完客户端向服务器的请求消息后，然后再了解服务器向客户端返回的响应消息。这种类型的消息也是由一个起始行，一个或者多个头信息，一个指示头信息结束的空行和可选的消息体组成。

HTTP 的头信息包括通用头、请求头、响应头和实体头 4 个部分。每个头信息由一个域名，冒号（:）和域值三部分组成。域名是大小写无关的，域值前可以添加任何数量的空格符。头信息可以被扩展为多行，在每行开始处，使用至少一个空格或制表符。

例如，爬虫程序发出 GET 请求：

```
GET /index.html HTTP/1.1
```

服务器返回响应：

```
HTTP /1.1 200 OK
Date: Apr 11 2006 15:32:08 GMT
Server: Apache/2.0.46(win32)
Content-Length: 119
Content-Type: text/html

<HTML>
<HEAD>
<LINK REL="stylesheet" HREF="index.css">
</HEAD>
<BODY>
<IMG SRC="image/logo.png">
</BODY>
</HTML>
```

GET 请求的头显示类似下面的信息：

```
GET / HTTP/1.0
 Host: www.lietu.com
 Connection: Keep-Alive
```

响应头显示类似如下信息：

```
HTTP/1.0 200 OK
 Date: Sun, 19 Mar 2006 19:39:05 GMT
 Content-Length: 65730
```

```
Content-Type: text/html
Expires: Sun, 19 Mar 2006 19:40:05 GMT
Cache-Control: max-age=60, private
Connection: keep-alive
Proxy-Connection: keep-alive
Server: Apache
Last-Modified: Sun, 19 Mar 2006 19:38:58 GMT
Vary: Accept-Encoding,User-Agent
Via: 1.1 webcache (NetCache NetApp/6.0.1P3)
```

服务器返回的响应第一行就包括状态码。状态码是一个由 3 个数字组成的结果代码。爬虫可以用状态码识别 Web 服务器处理的情况。状态码的第一个数字定义响应的类别，后两个数字有分类的作用，具体描述如下。

（1）1XX：信息响应类，表示接收到请求并且继续处理；

（2）2XX：处理成功响应类，表示动作被成功接收、理解和接受；

（3）3XX：重定向响应类，为了完成指定的动作，必须接受进一步处理；

（4）4XX：客户端错误，客户请求包含语法错误或者是不能正确执行；

（5）5XX：服务端错误，服务器不能正确执行一个正确的请求。

HTTP 常用状态码如表 3-2 所示。

表 3-2　HTTP 常用状态码

状态代码	代码描述	处理方式
200	请求成功	获得响应的内容，进行处理
201	请求完成，结果是创建了新资源。新创建资源的 URI 可在响应的实体中得到	爬虫中不会遇到
202	请求被接受，但处理尚未完成	阻塞等待
204	服务器端已经实现了请求，但是没有返回新的信息。如果客户是用户代理，则无须为此更新自身的文档视图	丢弃
300	该状态码不被 HTTP/1.0 的应用程序直接使用，只是作为 3XX 类型回应的默认解释。存在多个可用的被请求资源	若程序中能够处理，则进行进一步处理；如果程序中不能处理，则丢弃
301	请求到的资源都会分配一个永久的 URL，这样就可以在将来通过该 URL 来访问此资源	重定向到分配的 URL
302	请求到的资源在一个不同的 URL 处临时保存	重定向到临时的 URL
304	请求的资源未更新	丢弃
400	非法请求	丢弃

续表

状态代码	代码描述	处理方式
401	未授权	丢弃
403	禁止	丢弃
404	没有找到	丢弃
500	服务器内部错误	丢弃
502	错误网关	丢弃
503	服务器暂时不可用	丢弃

如下代码检查 requests 库中的状态代码。

```
import requests

url = "https://pytutorial.com/"

res = requests.get(url)

print(res.status_code)
```

HTTP 请求一张图片，如果没有数据能不能做出判断？看返回的 HTTP 状态码，没有找到就返回 404 状态码。有时候，也可能返回一个统一的 404 错误页面。

在提交表单的时候，如果不指定方法，则默认为 GET 请求，表单中提交的数据将会附加在 url 之后，以 "?" 与 url 分开。字母数字字符原样发送，但空格转换为 "+" 号，其他符号转换为 %XX，其中 XX 为该符号以十六进制表示的 ASCII 值。GET 请求提交的数据放置在 HTTP 请求协议头中，而 POST 提交的数据则放在实体数据中。GET 方式提交的数据最多只能有 1024 字节，而 POST 则没有此限制。

例如，程序发出 HEAD 请求：

```
HEAD /index.jsp HTTP/1.0
```

服务器返回响应：

```
HTTP/1.1 200 OK
Server: Apache-Coyote/1.1
Content-Type: text/html;charset=UTF-8
Content-Length: 5367
Date: Fri, 08 Apr 2011 11:08:24 GMT
Connection: close
```

3.5.2 HTML 文档

HTML 元素是由单个或一对标签定义的包含范围。一个标签就是左右分别有一个小于号（<）和大于号（>）的字符串。开始标签是指不以斜杠（/）开头的标签，其内是一串允许的属性/值对。结束标签则是以一个斜杠（/）开头的。

例如一个 HTML 元素 p：

```
<p>This is some text...</p>
```

<p> 是开始标签，</p> 是结束标签。

标签可以有属性，例如 img 标签：

```
<img src="logo.gif" alt="logo" />
```

img 标签中，属性 alt 的值是 "logo"。

一个标准的 HTML 文件应该以 <html> 开始标签开始文档，中间包含 <head> 与 <body> 等元素，其中 <head> 部分中可以定义页面的标题、简介、编码格式等内容，<body> 部分为在浏览器中显示的页面正文。下面的代码为一个不包含内容的标准 HTML 文档结构：

```
<html>
    <head>
    </head>
    <body>
    </body>
</html>
```

字符引用和实体引用都是以一个和号（&）开始并以一个分号（;）结束。如果使用的是字符引用，需要在和号之后加上一个井号（#），之后是所需字符的十进制代码或十六进制代码。如果使用的是实体引用，在和号之后写上字符的助记符。

3.5.3 使用 curl 命令下载网络资源

curl 命令的语法为：

```
curl [options] [URL]
```

如果没有任何命令行参数，curl 命令将获取一个文件并将其内容显示到标准输出。

```
curl https://www.digitalocean.com/robots.txt
```

要将远程文件保存到本地系统，并使文件名与下载服务器的文件名相同，请添加 --

remote-name 参数，或使用 -O 选项：

```
curl -O https://www.digitalocean.com/robots.txt
```

curl 不显示文件的内容，而是显示一个基于文本的进度表，并将文件保存为与远程文件名相同的名称。

该文件包含与之前看到的内容相同的内容：

```
Output
User-agent: *
Disallow:

sitemap: https://www.digitalocean.com/sitemap.xml
sitemap: https://www.digitalocean.com/community/main_sitemap.xml.gz
sitemap: https://www.digitalocean.com/community/questions_sitemap.xml.gz
sitemap: https://www.digitalocean.com/community/users_sitemap.xml.gz
```

现在，让我们看看为下载的文件指定一个文件名。

可能已经有一个本地文件与远程服务器上的文件同名。

为了避免覆盖同名的本地文件，请使用 -o 或 --output 参数，后跟要保存内容的本地文件的名称。

执行以下命令下载远程 robots.txt 文件到本地名为 do-bots.txt 的文件：

```
curl -o do-bots.txt https://www.digitalocean.com/robots.txt
```

默认情况下，curl 不跟随重定向，因此当文件移动时，可能无法获得预期的结果。让我们看看如何解决这个问题。

到目前为止，所有示例都包含了包含 https:// 协议的完全限定 URL。如果你碰巧去拿 robots.txt 文件，且仅指定 www.digitalocean.com，则不会看到任何输出，因为 DigitalOcean 将请求从 http:// 重定向到 https://。

可以使用 -I 标志来验证这一点，该标志显示请求头而不是文件的内容：

```
curl -I www.digitalocean.com/robots.txt
```

输出显示 URL 已重定向。输出的第一行显示它被移动了，位置行显示移动到了哪里：

```
HTTP/1.1 301 Moved Permanently
Cache-Control: max-age=3600
Cf-Ray: 65dd51678fd93ff7-YYZ
Cf-Request-Id: 0a9e3134b500003ff72b9d0000000001
Connection: keep-alive
Date: Fri, 11 Jun 2021 19:41:37 GMT
Expires: Fri, 11 Jun 2021 20:41:37 GMT
Location: https://www.digitalocean.com/robots.txt
```

```
Server: cloudflare
...
```

可以使用 curl 手动发出另一个请求，也可以使用 --location 或 -L 参数，该参数告诉 curl 在遇到重定向时将请求重做到新位置，试试看：

```
curl -L www.digitalocean.com/robots.txt
```

这一次可以看到输出了，因为 curl 跟随了重定向。

您可以将 -L 参数与前面提到的一些参数结合起来，将文件下载到本地系统：

```
curl -L -o do-bots.txt www.digitalocean.com/robots.txt
```

将数据传输速率限制为 1kb/s：

```
curl http://www.tutorialspoint.com/unix/ --limit-rate 1k -o unix.html
```

通过代理服务器下载：

```
curl -x proxy.example.com:3128 http://www.tutorialspoint.com/unix/
```

3.5.4 使用 wget 命令下载网页

wget 是一个免费实用程序，用于从 Web 以非交互式的方式下载文件。它支持 HTTP、HTTPS 和 FTP 协议，以及通过 HTTP 代理进行检索。

默认情况下，很容易调用 wget，基本语法为：

```
wget [option]… [URL]…
```

wget 下载命令行上指定的所有 URL。

可以使用 wget 命令递归地下载 HTTP 服务器某个目录下的所有文件，例如，递归下载所有在"ddd"目录下的文件：

```
#wget -r -np -nH --cut-dirs=3 -R index.html http://hostname/aaa/bbb/ccc/ddd/
```

使用到的选项解释如下。

（1）recursively (-r)，递归遍历；
（2）不到上级目录，例如 ccc/… (-np)；
（3）不保存文件到域名文件夹 (-nH)；
（4）除了 ddd，忽略前 3 个文件夹 aaa、bbb、ccc (--cut-dirs=3)；
（5）不包含 index.html 文件 (-R index.html)。

可以使用 wget -i 命令下载多个文件。首先创建一个文本文件，放入需要下载的 url，例如 download.txt

```
http://roll.mil.news.sina.com.cn/col/zgjq/index1.shtml
http://roll.mil.news.sina.com.cn/col/zgjq/index2.shtml
```

然后 wget -i 命令下载这些网页：

```
$ wget -i download.txt
```

可以添加很多 url 到这个文本文件，也可以只下载网站中的图像文件。

```
$ wget -i download.txt -r -P ./ -A jpg,jpeg,gif,png,bmp
```

可以用参数 t 限定重试次数。

```
wget -t 15 -i t_shop_data1.txt -r -P ./ -A jpg,jpeg,gif,png,bmp
```

参数 T 设置超时时间：

```
wget -t 15 -T 5 -i t_shop_data1.txt -r -P ./ -A jpg,jpeg,gif,png,bmp
```

也可以通过 wait 参数设置总的等待时间：

```
wget -r -p -k -nc -e robots=off --wait 1.0 -A bmp,png,jpg,jpeg,gif -i t_shop_data1.txt
```

3.5.5 下载静态网页

爬虫程序向服务器 www.example.com 发出 GET 请求请求根目录下的网页：

```
GET / HTTP/1.1
```

服务器返回响应：

```
HTTP/1.1 200 OK
Accept-Ranges: bytes
Age: 532288
Cache-Control: max-age=604800
Content-Type: text/html; charset=UTF-8
Date: Thu, 09 Mar 2023 08:16:32 GMT
Etag: "3147526947"
Expires: Thu, 16 Mar 2023 08:16:32 GMT
Last-Modified: Thu, 17 Oct 2019 07:18:26 GMT
Server: ECS (sab/5798)
Vary: Accept-Encoding
X-Cache: HIT
Content-Length: 1256

<!doctype html>
<html>
<head>
```

```html
        <title>Example Domain</title>

        <meta charset="utf-8" />
        <meta http-equiv="Content-type" content="text/html; charset=utf-8" />
        <meta name="viewport" content="width=device-width, initial-scale=1" />
        <style type="text/css">
        body {
            background-color: #f0f0f2;
            margin: 0;
            padding: 0;
                font-family: -apple-system, system-ui, BlinkMacSystemFont, "Segoe UI", "Open Sans", "Helvetica Neue", Helvetica, Arial, sans-serif;

        }
        div {
            width: 600px;
            margin: 5em auto;
            padding: 2em;
            background-color: #fdfdff;
            border-radius: 0.5em;
            box-shadow: 2px 3px 7px 2px rgba(0,0,0,0.02);
        }
        a:link, a:visited {
            color: #38488f;
            text-decoration: none;
        }
        @media (max-width: 700px) {
            div {
                margin: 0 auto;
                width: auto;
            }
        }
        </style>
</head>

<body>
<div>
    <h1>Example Domain</h1>
    <p>This domain is for use in illustrative examples in documents. You may use this domain in literature without prior coordination or asking for permission.</p>
    <p><a href="https://www.iana.org/domains/example">More information...</a></p>
</div>
</body>
</html>
```

服务器返回的响应中包含了根目录下的网页内容。

HTTP 协议使用了面向连接的 TCP 协议。HTTP 协议本身是无状态的。当请求一个网页，服务器返回页面之后，这个连接就没有了。如果需要记录用户的登录状态，可以使用 Cookie。

Socket 是一个端到端的通信管道。IP 地址和端口号组成了 Socket 地址。Socket 地址有两类：局域 Socket 地址和远程 Socket 地址。只有 TCP 协议才有远程 Socket 地址。服务器为每个客户端创建一个 Socket，这些 Socket 分享同样的局域 Socket 地址。Socket 使用的传输协议可以是 TCP 或者 UDP 等。

下载网页时，爬虫客户端和 Web 服务器端建立 Socket 连接。Web 服务器默认使用 80 端口。

```
import socket

sock = socket.socket(socket.AF_INET, socket.SOCK_STREAM)
sock.connect(("www.example.com", 80))
sock.send(b"GET / HTTP/1.1\r\nHost:www.example.com\r\n\r\n")
response = sock.recv(4096)
sock.close()
print(response.decode())
```

利用 requests 库可以很轻易地根据给定 URL 地址下载网页。

```
html = requests.get(url).text
```

除了网页源代码，还可以打印返回的头信息：

```
import requests
getdata = requests.get('https://jsonplaceholder.typicode.com/users')
print(getdata.headers)
```

虽然大部分网络资源使用 HTTP 协议下载，但是为了加密通信内容及鉴定 Web 服务器的身份，还有 HTTPS 协议。如果使用 https:// 来访问某个网站，就表示此网站部署了 SSL 证书。

使用 requests 库时，不需要对 HTTPS 请求进行特殊处理。

为了节约带宽，有的网页是用 GZIP 流传输过来的。把得到的 GZIP 流内容保存到文件的示例：

```
import requests

url = "http://news.sohu.com/20110705/n312398236.shtml"

r = requests.get(url, stream=True)
```

```
print(r.headers.get('Content-Encoding'))
local_filename = "test.gzip";
with open(local_filename, 'wb') as f:
    for chunk in r.raw.stream(1024, decode_content=False):
        if chunk:
            f.write(chunk)
```

3.5.6　使用 Selenium 下载动态内容

因为网页中可能包括 JavaScript，所以对于这样的动态网页需要特别处理，例如有些新闻评论是动态页面，也就是说内容是用 JavaScript 生成的，对于这样的内容，需要 JavaScript 渲染以后才能得到。

Selenium 是一个开源的伞式项目，包含一系列旨在支持浏览器自动化的工具和库。它支持包括 Python 在内的几种流行编程语言的绑定。

Selenium 使用 Webdriver 协议来自动化各种流行浏览器（如 Firefox、Chrome 和 Safari）上的进程。

首先安装 Selenium：

```
pip install selenium
```

为了使用 Selenium 自动化操作浏览器，需要浏览器对应的驱动。Selenium FirefoxDriver 或 Selenium GeckoDriver 可以从 Mozilla 的官方 GitHub 存储库中（https://github.com/mozilla/geckodriver/releases）下载。

使用 FireFox 驱动的示例如下：

```
from selenium import webdriver
from selenium.webdriver.common.keys import Keys
from selenium.webdriver.common.by import By
from selenium.webdriver.chrome.service import Service

s = Service(r"D:/soft/geckodriver-v0.31.0-win64/geckodriver.exe")
driver = webdriver.Firefox(service=s)

driver.get("http://www.python.org")
assert "Python" in driver.title
elem = driver.find_element(By.NAME, "q")
elem.clear()
elem.send_keys("pycon")
elem.send_keys(Keys.RETURN)
assert "No results found." not in driver.page_source
```

```
driver.close()
```

为了简单起见，可以使用名为 Webdriver Manager 的第三方 Python 库来获取正确的驱动程序并对其进行配置。

安装 webdriver-manager：

```
pip install webdriver-manager
```

让 Firefox 以无头方式运行的代码如下：

```
from selenium import webdriver
from selenium.webdriver.chrome.service import Service
from selenium.webdriver.firefox.options import Options
from webdriver_manager.firefox import GeckoDriverManager

try:
    s = Service(GeckoDriverManager(cache_valid_range=1).install())
    options = Options()
    options.headless = True
    brower = webdriver.Firefox(service=s, options=options)

    brower.get('https://pythonbasics.org')
    print(brower.page_source)
finally:
    try:
        brower.close()
    except:
        pass
```

运行 JavaScript 脚本：

```
from selenium import webdriver
from selenium.webdriver.chrome.service import Service
from webdriver_manager.firefox import GeckoDriverManager

s = Service(GeckoDriverManager(cache_valid_range=1).install())

driver = webdriver.Firefox(service=s)

driver.execute_script("alert('running javascript');")
```

通过执行 JavaScript 脚本来实现页面向下滚动：

```
import time

from selenium import webdriver
from selenium.webdriver.chrome.service import Service
```

```
from webdriver_manager.firefox import GeckoDriverManager

s = Service(GeckoDriverManager(cache_valid_range=1).install())

driver = webdriver.Firefox(service=s)

driver.get('https://pythonbasics.org')

totalHeight = driver.execute_script("return document.body.scrollHeight")

scrollNum: int = int(totalHeight / 300.0)  # 滚动次数

for i in range(scrollNum):
        driver.execute_script("window.scrollBy(0,document.body.scrollHeight/" + str(scrollNum) + ")")
        time.sleep(2)

driver.close()
```

可以设置等待条件等待需要的时间长度：

```
from selenium import webdriver
from selenium.webdriver.chrome.service import Service
from selenium.webdriver.support import expected_conditions
from selenium.webdriver.support.wait import WebDriverWait
from webdriver_manager.firefox import GeckoDriverManager

s = Service(GeckoDriverManager(cache_valid_range=1).install())

driver = webdriver.Firefox(service=s)

driver.get("https://chercher.tech/practice/explicit-wait")
wait = WebDriverWait(driver, 3)
wait.until(expected_conditions.title_contains("Practice"))

driver.quit()
```

3.5.7 重试

为了提高爬虫的稳定性，往往在放弃下载之前多次重试。

向 HTTP 客户端添加重试策略很简单。我们创建一个 HTTPAdapter 并将我们的策略传递给适配器。

```
import requests
```

```
from requests.adapters import HTTPAdapter, Retry

retry_strategy = Retry(
    total=3,
    status_forcelist=[429, 500, 502, 503, 504],
    allowed_methods=["HEAD", "GET", "OPTIONS"]
)
adapter = HTTPAdapter(max_retries=retry_strategy)
http = requests.Session()
http.mount("https://", adapter)
http.mount("http://", adapter)

response = http.get("https://api.twilio.com/")
```

默认的 Retry 类提供了合理的默认值，高度可配置，这里是最常见参数的简要说明。下面的参数包括 requests 库使用的默认参数。

`total=10`

重试尝试的总数。如果失败请求或重定向的数量超过此数量，客户端将抛出 urllib3.exceptions.MaxRetryError 异常。

`status_forcelist=[413, 429, 503]`

要重试的 HTTP 响应代码。您可能希望重试常见的服务器错误（500、502、503、504），因为服务器和反向代理并不总是遵守 HTTP 规范。

`allowed_methods=["HEAD", "GET", "OPTIONS"]`

要重试的 HTTP 方法。默认情况下，这包括除 POST 之外的所有 HTTP 方法，因为 POST 可能导致新的插入。

`backoff_factor=0`

它允许您更改进程在失败的两次请求之间休眠的时间，算法如下：

`{backoff factor} * (2 ** ({number of total retries} - 1))`

例如，如果退避因子设置为：

（1）1 秒，相继的睡眠将是 0.5、1、2、4、8、16、32、64、128、256

（2）2 秒，相继的睡眠时间将是 -1、2、4、8、16、32、64、128、256、512

（3）10 秒，相继的睡眠时间将是 -5、10、20、40、80、160、320、640、1280、2560

该值呈指数增长，这是重试策略的合理默认实现。

3.6 下载图片

如下代码把 URL 地址中的图片写到文件中去。

```
import requests

f = open('NASA3.jpg','wb')
response = requests.get('http://www.python.org/images/success/nasa.jpg')
f.write(response.content)
f.close()

print("download successful")
```

3.7 网络爬虫的遍历与实现

互联网中有海量网页信息，它们是通过超级链接进行相互跳转的，这些超级链接把网页组成了一张很大的网，如图 3-7 所示。网络爬虫的抓取原理就是从互联网中的一个网页开始，根据网页中的超级链接逐个抓取网页中链接的其他网页。

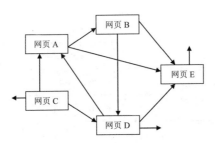

图 3-7 互联网网页链接图

网页通过超级链接相互链接，组成了一个庞大的无形的网，信息量十分庞大，网络爬虫不可能抓取所有的网页信息。所以，使用网络爬虫抓取网页要遵循一定的原则，主要有广度优先原则和最佳优先原则。

广度优先是指网络爬虫会先抓取起始网页中链接的所有网页，然后再选择其中的一个链接网页，继续抓取在此网页中链接的所有网页。这是最常用的方式，这个方法也可以让网络爬虫并行处理，提高其抓取速度。以图 3-8 中的图为例说明广度遍历的过程。

例如在图 3-8 中，A 为种子节点，则首先遍历 A（第一层），接着是 BCDEF（第二层），接着遍历 GH（第三层），最后遍历 I（第四层）。

广度优先遍历使用一个队列来实现 Todo 表，先访问的网页先扩展。针对图 3-8，广度优先遍历的执行过程如表 3-3 所示。

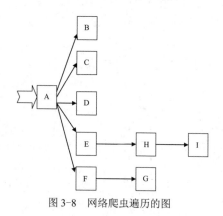

图 3-8　网络爬虫遍历的图

表 3-3　广度优先遍历过程表

Todo 队列	Visited 集合
a	null
b c d e f	a
c d e f	a b
d e f	a b c
e f	a b c d
f h	a b c d e
h g	a b c d e f
g i	a b c d e f h
i	a b c d e f h g
null	a b c d e f h g i

庄子曾说："吾生也有涯，而知也无涯，以有涯随无涯，殆已"。在学习和工作的时候，需要分辨事情的轻重缓急，否则一味蛮干，最终结果只能是"殆已"。对于浩瀚无边的互联网而言，网络爬虫涉及页面确实只是冰山一角，因此，需要以最小的代价（硬件、带宽）获取到最大的利益（数量最多的重要的网页）。

为了先抓取重要的网页，可以采用最佳优先。最佳优先爬虫策略也称为"页面选择问题"（PageSelection），通常，这样保证在有限带宽条件下，尽可能地照顾到重要性高的网页。

如何实现最佳优先爬虫呢？最简单的方式可以使用优先级队列（PriorityQueue）来实现 Todo 表，这样，每次选出来扩展的 URL 就是具有最高重要性的网页。在队列中，先进入的元素先出，但是在优先队列中，优先级高的元素先出队列。

比如，假设上图的节点重要性 D>B>C>A>E>F>I>H，则整个遍历过程如表 3-4 所示。

表 3-4 最佳优先遍历过程表

Todo 优先级队	Visited 集合
A	null
B,C,D,E,F	A
B,C,E,F	A,D
C,E,F	A,B,D
E,F	A,B,C,D
F,H	A,B,C,D,E
H,G	A,B,C,D,E,F
H	A,B,C,D,E,F,G
I	A,B,C,D,E,F,H,G
null	A,B,C,D,E,F,H,I

使用 Python 中的优先队列：

```
import queue

# Initializing a priority queue
pqueue = queue.PriorityQueue()

# Using put() function to insert elements
pqueue.put((4,'China'))
pqueue.put((1,'Russia'))
pqueue.put((2,'England'))
pqueue.put((5,'Nepal'))
pqueue.put((3,'Italy'))

# Get 3 lowest priority countries
print(pqueue.get())
print(pqueue.get())
print(pqueue.get())

# Return total number of elements
print(pqueue.qsize())

# Use the full() function to check priority queue is full or not.
print(pqueue.full())
```

```
# Use the empty() function to check priority queue is empty or not.
print(pqueue.empty())
```

3.8 robots 协议

网络爬虫要遵循 robots 协议，也就是要读取 robots.txt 文件。

robots.txt 文件基本上是一种指定爬虫访问策略的方法，可以通过 HTTP 在本地 URL "/robots.txt" 上访问该文件。选择这种方法是因为它可以很容易地在任何现有的 WWW 服务器上实现，并且爬虫可以通过单个文档检索找到访问策略。

urllib.robotparser 提供了 RobotFileParser 类，它回答了有关特定用户代理是否可以在发布了 robots.txt 文件的网站上获取 URL 的问题。

示例代码如下：

```
import urllib.robotparser

parser = urllib.robotparser.RobotFileParser()
parser.set_url("https://cppsecrets.com/robots.txt")
parser.read()
print(parser.crawl_delay("*"))    # 输出 None
print(parser.can_fetch("*", "https://www.cppsecrets.com/")) # 输出 True
```

为了方便爬虫遍历和更新网站内容，网站可以设置 Sitemap.xml。Sitemap.xml 也就是网站地图，不过这个网站地图是用 XML 写的，例如：

https://search.gov/sitemap.xml

在其中列出网站中的网址以及关于每个网址的其他元数据（上次更新的时间、更改的频率以及相对于网站上其他网址的重要程度等），以便搜索引擎抓取网站。

完整格式如下：

```
<?xml version="1.0" encoding="UTF-8"?>
<urlset>
    <url>
            <loc>https://example.gov/blog/file1.html</loc>
            <lastmod>2021-07-17</lastmod>
            <changefreq>daily</changefreq>
            <priority>1.0</priority>
    </url>
    <url>
            <loc>https://example.gov/policy/new-policy.html</loc>
```

```
            <lastmod>2021-07-17</lastmod>
            <changefreq>weekly</changefreq>
            <priority>0.9</priority>
        </url>
        ...
</urlset>
```

其中的 XML 标签的含义说明如下。

（1）loc：页面永久链接地址。

（2）lastmod：页面最后的修改时间。

（3）changefreq：页面内容的更新频率。

（4）priority：相对于其他页面的优先权。

对于有网站地图的网站，爬虫可以利用这个网站地图遍历网站和增量抓取。

以下是有关如何读入 XML 格式的网站地图文件的控制台应用程序。

```
from usp.tree import sitemap_tree_for_homepage

tree = sitemap_tree_for_homepage('https://seoagilitytools.com')
print(tree)
# all_pages() returns an Iterator
for page in tree.all_pages():
    print(page)
```

3.9 连接池

当一个新的连接请求进来的时候，连接池管理器检查连接池中是否包含任何没用的连接，如果有的话，就返回一个。

如果连接池中所有的连接都忙并且最大的连接池数量没有达到，就创建新的连接并且增加到连接池。当连接池中在用的连接达到最大值，所有的新连接请求进入队列，直到一个连接可用或者连接请求超时。

连接池包含如下参数。

（1）连接超时：控制请求一个新连接的等待时间，如果超时，将会抛出一个异常。

（2）最大连接数：声明连接池的最大值，缺省是 100。

（3）最小连接数：连接池创建时的初始连接数量。

程序一开始初始化创建若干数量的长链接，给他们设置一个标识位，这个标识位表示该链接是否空闲的状态。当需要发送数据的时候，系统给它分配一个当前空闲的链接。同时，将得到的链接设置为"忙"，当数据发送完毕后，把链接标识位设置为"闲"，让系统可以分配给下个用户。

Session 对象允许跨请求保留某些参数，它还在从 Session 实例发出的所有请求中保留 cookie，并将使用 urllib3 的连接池。因此，如果对同一主机发出多个请求，则底层 TCP 连接将被重用，这可能会导致性能显著提高。

让我们通过将 cookie 设置给 URL，然后再次发出请求以检查 cookie 是否已设置来说明会话对象的使用。

```
# import requests module
import requests

# create a session object
s = requests.Session()

# make a get request
s.get('https://httpbin.org/cookies/set/sessioncookie/123456789')

# again make a get request
r = s.get('https://httpbin.org/cookies')

# check if cookie is still set
print(r.text)
```

3.10 URL 地址查新

在科技论文发表时，为了避免重复研究和抄袭，需要到专门的科技情报所做论文查新。为了避免重复抓取，URL 地址也需要查新。判断解析出的 URL 是否已经遍历过叫作 URLSeen 测试。URLSeen 测试对爬虫性能有重要的影响。本节介绍两种实现快速 URLSeen 测试的方法。

在介绍爬虫架构的时候，我们讲解了 Frontier 组件的作用。它作为一个基础的组件，为爬虫提供 URL。因此，在 Frontier 中有一个数据结构来存储 URL。在一些小的爬虫程序中，使用内存队列（List、HashMap 或 Queue）或者优先级队列来存储 URL，但内存是有限的。通常在商业应用中，URL 地址数据量非常大。早期的爬虫经常把 URL 地址放在数据库表中，但数据库对于这种简单的结构化存储来说效率太低，可以用 B+ 树（https://github.com/NicolasLM/bplustree），还可以考虑使用内存数据结构存储 Redis 来存储。

3.10.1 Redis 数据库

Redis 的强大之处在于多种类型的数据结构的可用性。每一个数据结构都支持以特定方式表示数据，这有助于提高访问速度。Redis 的数据结构有：

（1）Strings；

（2）Hashes；

（3）Lists；

（4）Sets；

（5）Sorted sets；

（6）Bitmaps；

（7）Hyperlogs；

（8）Geo-spatial indexes；

（9）Streams。

Windows 下的 Redis 可从 https://github.com/zkteco-home/redis-windows 下载。

启动 Redis 服务：

```
>redis-server.exe
```

安装 Redis Python 客户端：

```
>pip install redis
```

使用 Redis Python API 集合操作的完整的示例代码如下：

```
# Example Python program to demonstrate set membership operation in Redis
import redis

redisClient = redis.StrictRedis(host='localhost',
                                port=6379,
                                db=0)

# Define a set of urls
redisClient.sadd("urlSeen", "https://seoagilitytools.com", "https://httpbin.org")

# Check if https://httpbin.org is seen
print(redisClient.sismember("urlSeen", "https://httpbin.org"))

# Check if https://httpbin.org/test is seen
print(redisClient.sismember("urlSeen", "https://httpbin.org/test"))
```

使用 Redis 实现的 URL 地址查询方法如下：

```
import redis
```

```
class UrlSeenDetector:
    def __init__(self):
        self.redisClient = redis.StrictRedis(host='localhost',
                                             port=6379,
                                             db=0)

    def detect(self, url):
        ret = self.redisClient.sismember("urlSeen", url)
        if ret:
            return ret
        self.redisClient.sadd("urlSeen", url)
```

3.10.2 布隆过滤器

判断 URL 地址是否已经抓取过还可以借助于布隆过滤器（Bloom Filter）。布隆过滤器的实现方法是：利用内存中的一个长度是 m 的位数组 B，对其中所有位都置 0，如图 3-9 所示。

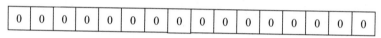

图 3-9　位数组 B 的初始状态

然后对每个遍历过的 URL 根据 k 个不同的散列函数执行散列，每次散列的结果都是不大于 m 的一个整数 a。根据散列得到的数在位数组 B 对应的位上置 1，也就是让 B[a]=1。图 3-10 所示显示了放入 3 个 URL 后位数组 B 的状态，这里 k=3。

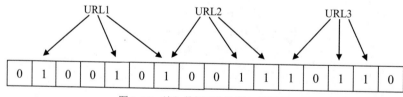

图 3-10　放入数据后位数组 B 的状态

每次插入一个 URL，也执行 k 次散列，只有当全部位都已经置 1 了才认为这个 URL 已经遍历过。bloom-filter（https://github.com/remram44/python-bloom-filter）实现了布隆过滤器。如下是使用布隆过滤器的一个示例：

```
from bloom_filter2 import BloomFilter

# instantiate BloomFilter with custom settings,
# max_elements is how many elements you expect the filter to hold.
```

```
# error_rate defines accuracy; You can use defaults with
# `BloomFilter()` without any arguments. Following example
# is same as defaults:
bloom = BloomFilter(max_elements=10000, error_rate=0.1)

# Mark the key as seen
bloom.add("http://www.python.org")
bloom.add("https://httpbin.org")

# Now check
assert "https://httpbin.org" in bloom
```

布隆过滤器如果返回不包含某个项目，那肯定就是没往里面增加过这个项目，如果返回包含某个项目，但其实可能没有增加过这个项目，所以有误判的可能。对爬虫来说，使用布隆过滤器的后果是可能导致漏抓网页。如果想知道需要使用多少位才能降低错误概率，可以从表 3-5 的存储项目和位数比率估计布隆过滤器的误判率。

表 3-5 布隆过滤器误判率表

比率（items:bits）	误 判 率
1:1	0.63212055882856
1:2	0.39957640089373
1:4	0.14689159766038
1:8	0.02157714146322
1:16	0.00046557303372
1:32	0.00000021167340
1:64	0.00000000000004

为每个 URL 分配两个字节就可以达到千分之几的冲突，例如一个比较保守的实现，为每个 URL 分配了 4 个字节，项目和位数比是 1:32，误判率是 0.00000021167340。对于 5000 万数量级的 URL，布隆过滤器只占用了 200M 的空间，并且排重速度超快，一遍下来不到两分钟。

用一个类封装 BloomFilter。

```
from bloom_filter2 import BloomFilter

class UrlFilter:
    def __init__(self):
        self.bloom = BloomFilter(max_elements=3000000, error_rate=0.001)
```

```
    def contain_url(self, url):
        if url in self.bloom:
            return True
        self.bloom.add(url)
        return False
```

3.11 抓取 RSS

因为 XML 比 HTML 更规范，所以出现了 XML 格式封装的数据源。RSS 是对网站栏目的一种 XML 格式的封装。一些博客或者新闻网站提供了 RSS（Really Simple Syndication）格式的输出，方便程序快速访问更新的信息。每一条信息叫作 Feed。RSS 抓取的第一步是解析 RSS 数据源，例如，https://arminreiter.com/feed/ 就是一个 RSS 数据源。

为了读取 RSS 种子，首先安装包 feedparser，然后使用 feedparser.parse() 方法读入 RSS 种子。

```
import feedparser
NewsFeed = feedparser.parse("https://arminreiter.com/feed/")
entry = NewsFeed.entries[1]

print (entry.keys())
```

3.12 网页更新

经常有人会问："有没有什么新消息？"，这说明人的大脑是增量获取信息的，对爬虫来说也是如此。网站中的内容经常会变化，这些变化经常在网站首页或者目录页有反应。为了提高采集效率，往往考虑增量采集网页，可以把这个问题看成是被采集的 Web 服务器和存储库同步的问题。更新网页内容的基本原理是：下载网页时，记录网页下载时的时间，增量采集这个网页时，判断 URL 地址对应的网页是否有更新。

HTTP 1.1 声明支持一种特殊类型的 HTTP Get，叫作 HTTP 条件 Get。如果文件在某个条件下没有修改，则 HTTP 条件 Get 不下载这个网页。判断网页是否修改的方法包括：If-Modified-Since、If-Unmodified-Since、If-Match、If-None-Match 或者 If-Range 头信息。

爬虫发送条件 GET 请求：

```
GET / HTTP/1.1
Host: www.example.com:80
If-Modified-Since:Thu, 4 Feb 2010 20:39:13 GMT
```

```
Connection: Close
```

当没有更新时服务器的响应：

```
HTTP/1.0 304 Not Modified
Date: Thu, 04 Feb 2010 12:38:41 GMT
Content-Type: text/html
Expires: Thu, 04 Feb 2010 12:39:41 GMT
Last-Modified: Thu, 04 Feb 2010 12:29:04 GMT
Age: 28
Connection: close
```

如果服务器网页已经更新就会把客户端的请求当作一个普通的 Get 请求发送网页内容：

```
HTTP/1.0 200 OK
Date: Thu, 04 Feb 2010 12:49:46 GMT
Server: Apache
Last-Modified: Thu, 04 Feb 2010 12:49:05 GMT
Accept-Ranges: bytes
Cache-Control: max-age=60
Expires: Thu, 04 Feb 2010 12:50:46 GMT
Vary: Accept-Encoding
X-UA-Compatible: IE=EmulateIE7
Content-Length: 452785
Content-Type: text/html
Age: 11
Connection: close
/*....... 网页内容 ....... */
```

条件下载命令可以根据时间条件下载网页。再次请求已经抓取过的页面时，爬虫往 Web 服务器发送 If-Modified-Since 请求头，其中包含的时间是先前服务器端发过来的 Last-Modified 最后修改时间戳，这样让 Web 服务器端进行验证，通过这个时间戳判断爬虫上次抓过的页面是否有修改。如果有修改，则返回 HTTP 状态码 200 和新的内容；如果没有变化，则只返回 HTTP 状态码 304，告诉爬虫页面没有变化。这样可以大大减少在网络上传输的数据，同时也减轻了被抓取的服务器的负担。

看一下 HTTP 的 Get 代码示例。Response 对象中的头信息返回 ETag 和最后修改日期。

```
import requests

response = requests.get('http://www.python.org/images/success/nasa.jpg')

print(response.headers)
```

```
eTag = response.headers.get("ETag")
print(eTag)

lastModified = response.headers.get("Last-Modified")

print(lastModified)
```

看一下 HTTP 条件 Get 的代码示例，在取得 HTTP 响应前，它使用 Etag 去设置 If-None-Match 头域信息并且使用最后修改时间去设置 If-Modified-Since 头域信息。如果网页没有修改，状态码返回 304。

```
import feedparser

d = feedparser.parse('https://arminreiter.com/feed/')
d2 = feedparser.parse('https://arminreiter.com/feed/', etag=d.etag)

print(d2.status)

d3 = feedparser.parse('https://arminreiter.com/feed/', etag=d.etag, modified=d.modified)

print(d3.status)
```

3.13 进度条

爬取时往往需要显示抓取进度。使用 tqdm 库，我们可以制作控制台行进度条和带有 GUI 的进度条。通过利用这些进度条，可以查看我们是否卡在某个地方并立即着手处理。

安装 tqdm 库：

```
>pip install tqdm
>pip install ipywidgets
```

通过 range() 函数使用 tqdm：

```
from tqdm import tqdm
from time import sleep

for i in tqdm(range(10)):
    sleep(0.25)
```

把 tqdm 集成进爬虫：

```
from time import sleep
from tqdm import tqdm
```

```
for num in tqdm(range(1, 23)):
    url = f"https://slickdeals.net/computer-deals/?page={num}"
    print(url)
    sleep(0.25)
```

3.14 垂直行业抓取

采集一些新闻放入网站内容管理系统数据库,首先确定要采集的目录页首页,然后可以通过翻页遍历所有的目录页,提取每个目录页对应的详细页面。

以新浪新闻为例,同一个目录下的 URL 是:

```
http://roll.news.sina.com.cn/news/gjxw/hqqw/index.shtml
http://roll.news.sina.com.cn/news/gjxw/hqqw/index_2.shtml
http://roll.news.sina.com.cn/news/gjxw/hqqw/index_3.shtml
```

所有的目录页都符合

```
http://roll.news.sina.com.cn/news/gjxw/hqqw/index_XXX.shtml
```

这样的规律。其中 XXX 是页码号,页码号从 1 开始。使用循环生成 99 个目录页网址。

```
for pageNo in range(1, 100):
    url = f"http://roll.news.sina.com.cn/news/gjxw/hqqw/index_{pageNo}.shtml"
    print(url)
```

使用 BeautifulSoup 库提取目录页中的详细页面信息。BeautifulSoup 把网页转换成由节点组成的树。HTML 文档是 bs4.BeautifulSoup 类型,例如把所有的超文本链接打印出来:

```
from bs4 import BeautifulSoup
import requests
url = "http://roll.news.sina.com.cn/news/gjxw/hqqw/index_3.shtml"
req = requests.get(url)
req.encoding = req.apparent_encoding
soup = BeautifulSoup(req.text, "html.parser")

for tag in soup.find_all('a'):
    print(tag.get('href'))
```

提取出详细页面的链接后,再用类似的方法提取详细页中的新闻标题和正文,并放入数据库。这里使用 pyodbc 库与 Access 数据库进行交互。

需要指定数据库驱动程序的值是 {Microsoft Access Driver (*.mdb, *.accdb)}。指定数据源(Data Source)的值是一个文件名,例如 test_database.accdb。数据库连接参数通过一个

字符串指定,指定的多个值用分号分隔开,代码如下:

```
conn = pyodbc.connect(
    r'Driver={Microsoft Access Driver (*.mdb, *.accdb)};DBQ=d:\Test\test_database.accdb;')
```

创建表的 SQL 语句格式如下:

```
CREATE TABLE 表名称
(
列名称1 数据类型,
列名称2 数据类型,
列名称3 数据类型,
....
)
```

例如,创建一个 article 表用来存放文章:

```
CREATE TABLE article(
url   text
)
```

用 INSERT 语句把数据放入数据库:

```
INSERT INTO article (url) VALUES (' https://news.sina.com.cn/o/2017-10-09/doc-ifymrqmq1572406.shtml ')
```

在 Python 中,可以通过 Cursor 对象执行 SQL 语句。

```
import pyodbc

conn = pyodbc.connect(
    r'Driver={Microsoft Access Driver (*.mdb, *.accdb)};DBQ=d:\Test\test_database.accdb;')
cursor = conn.cursor()
ret = cursor.execute('INSERT INTO article (url) VALUES (?)','https://news.sina.com.cn/o/2017-10-09/doc-ifymrqmq1572406.shtml')

print(ret.rowcount)

conn.commit()
```

3.15 抓取限制的应对方法

对爬虫不友好的网站有各种各样的限制抓取的方法,所以爬虫应对的方法也不同。从原理上来说，只要浏览器可以访问，爬虫应该也可以访问。

3.15.1 模拟浏览器访问

有些网站检查请求头，只正常应答和浏览器一样的请求头。先检查下浏览器发送的请求头，然后用程序模拟发送和浏览器一样的请求头。

可以在 Chrome 浏览器中的开发者工具中看到 Chrome 浏览器发送的头信息。

网站往往检查 User-Agent 的值，可以把 User-Agent 的值设置为：

```
" Mozilla/5.0 (Windows NT 10.0; rv:91.0) Gecko/20100101 Firefox/91.0"
```

完整代码如下：

```
import requests

url = "https://www.shellhacks.com"
headers = {
    'User-Agent': 'Mozilla/5.0 (Windows NT 10.0; rv:91.0) Gecko/20100101 Firefox/91.0'
}

response = requests.get(url, headers=headers)
```

3.15.2 使用代理 IP

有些网站对于同一个 IP 在一段时间内的访问次数有限制，可以使用 Socket 代理来更改请求的 IP，这时可以通过大量不同的 Socket 代理循环访问网站。

在 proxies 属性中指定代理，代码如下：

```
import requests
proxies = {
'http': 'http://localhost:8080'
}
res = requests.get('http://httpbin.org/', proxies=proxies)
print(res.status_code)
```

需要用户名和密码身份验证的代理以不同的方式配置，但是，差异并没有那么大，所需要的只是对上面的语法进行调整和更改，以适应用于身份验证的用户名和密码。下面是显示如何使用用户名和密码身份验证配置代理的代码。

```
proxies = { 'https' : 'https://user:password@proxyip:port' }
```

```
r = requests.get('https://url', proxies=proxies)
```

轮换代理的代码如下:

```
import requests

from itertools import cycle

list_proxy = ['socks5://Username:Password@IP1:20000',
              'socks5://Username:Password@IP2:20000',
              'socks5://Username:Password@IP3:20000',
              'socks5://Username:Password@IP4:20000',
              ]

proxy_cycle = cycle(list_proxy)

proxy = next(proxy_cycle)

for i in range(1, 10):
    proxy = next(proxy_cycle)
    print(proxy)
    proxies = {
        "http": proxy,
        "https": proxy
    }

    r = requests.get(url='https://ident.me/', proxies=proxies)
    print(r.text)
```

3.15.3 抓取需要登录的网页

有些页面可以直接打开,而有些页面必须登录之后才能打开。对浏览器来说,网站会发送状态信息给浏览器,然后浏览器会返回这个状态信息给网站,这个状态信息叫作Cookie。登录后抓取信息的代码如下:

```
import requests
import time

headers = {
    "User-Agent": "Mozilla/5.0 (Windows NT 6.1; WOW64) AppleWebKit/537.36 (KHTML, like Gecko) Chrome/63.0.3239.132 Safari/537.36 QIHU 360SE"}
session = requests.Session()
params = {"username": "username", "pwd": "password", "formhash": "FA0334B8A2"}
mysession = session.post("https://www.yaozh.com/login/", params)
```

```
print(mysession.cookies.get_dict())
time.sleep(3)
mysession = session.get("https://www.yaozh.com/member/")
print(mysession.text)
```

3.16 保存信息

需要把抓取下来的文本存入数据库，有时候需要保存网页图像，例如，网页图像经过公证后，有可能作为法庭使用的证据。

3.16.1 SQLite 数据库

可以把结构化数据存入 SQLite 数据库。SQLite 是一个嵌入式关系数据库引擎，它是一个独立的，无服务器，零配置和事务性 SQL 数据库引擎。SQLite 实现了 SQL 的大部分 SQL-92 标准，并没有专门的进程管理 SQLite 数据库文件。SQLite 引擎不是一个独立的进程，相反，它是静态或动态链接到应用程序。SQLite 数据库是一个普通的磁盘文件，可以位于目录层次结构中的任何位置。

可以使用 Dbeaver（https://github.com/dbeaver/dbeaver）管理 SQLite 数据库。

在 CentOS 下可以使用如下命令安装 sqlite 包：

```
#yum install -y gcc gcc-c++ zlib-devel gdbm-devel readline-devel openssl-devel libffi-devel sqlite-devel
```

SQLite 附带了 sqlite3 命令行实用程序，它可用于对数据库发出 SQL 命令。现在我们将使用 sqlite3 命令行工具来创建一个新数据库。

```
$ sqlite3 test.db
SQLite version 3.6.22
Enter ".help" for instructions
Enter SQL statements terminated with a ";"
```

我们为 sqlite3 工具提供了一个参数。test.db 是一个数据库名称，它是我们磁盘上的单个文件。如果它存在，则打开；如果没有，则创建它。

```
sqlite> .tables
sqlite> .exit
$ ls
test.db
```

.tables 命令提供 test.db 数据库中的表名列表。test.db 数据库中目前没有表。.exit 命令终止 sqlite3 命令行工具的交互式会话。UNIX 命令 ls 显示当前工作目录的内容，我们可以看到 test.db 文件，所有数据都将存储在此单个文件中。

SQLite3 可以使用 sqlite3 模块与 Python 集成。如下代码创建一个数据库表并用数据填充它。

```python
import sqlite3

conn = sqlite3.connect(':memory:')

print("Opened database successfully") ;

conn.execute('''CREATE TABLE news
        (url TEXT PRIMARY KEY     NOT NULL,
        title           TEXT    NOT NULL,
        body            TEXT    NOT NULL);''')
cursor = conn.cursor()

data_tuple = ('https://news.sina.com.cn/o/2017-10-09/doc-ifymrqmq1572406.shtml', 'title', 'body')

ret = cursor.execute('INSERT INTO news (url,title,body) VALUES (?,?,?)',data_tuple)

print(ret.rowcount)
```

3.16.2 MySQL 数据库

MySQL 是一个基于服务器的数据库管理系统。一台服务器可能包含多个数据库。

CREATE TABLE 语句用于在 MySQL 数据库中创建表。在这里，您需要指定表的名称和每列的定义（名称和数据类型）。

以下是在 MySQL 中创建表的语法：

```
CREATE TABLE table_name(
    column1 datatype,
    column2 datatype,
    column3 datatype,
    .....
    columnN datatype,
);
```

名为 execute() 的方法（在游标对象上调用）接受两个变量：

（1）一个字符串值，表示要执行的查询。

(2)可选的 args 参数,可以是元组、列表或字典,表示查询的参数(占位符的值)。它返回一个整数值,表示查询影响的行数。

下面的示例在数据库 MySQL 中创建一个名为 news 的表并插入数据。

```
# importing required libraries
import mysql.connector

conn = mysql.connector.connect(
    host="localhost",
    user="root",
    passwd="pwd",
    database='mysql'
)

print(conn)
#Creating a cursor object using the cursor() method
cursor = conn.cursor()

#Creating table as per requirement
sql ='''CREATE TABLE IF NOT EXISTS news(
    url             TEXT    NOT NULL,
    title           TEXT    NOT NULL,
    body            TEXT    NOT NULL
)'''
cursor.execute(sql)

data_tuple = ('https://news.sina.com.cn/o/2017-10-09/doc-ifymrqmq1572406.shtml','title', 'body')

cursor.execute('INSERT INTO news (url,title,body) VALUES (%s,%s,%s)',data_tuple)

conn.commit()

# Disconnecting from the server
conn.close()
```

3.16.3 MongoDB 数据库

首先介绍在 Linux 下安装 MongoDB,安装服务器端,然后介绍通过 MongoDB 驱动程序和 MongoDB 打交道。

首先下载 MongoDB:

```
curl -O https://fastdl.mongodb.org/linux/mongodb-linux-x86_64-3.4.4.tgz
```

解压缩：

```
tar -zxvf mongodb-linux-x86_64-3.4.4.tgz
```

在首次启动 MongoDB 之前，需要创建 mongod 进程写入数据的目录。默认情况下，mongod 会将数据写入 /data/db 目录。使用以下命令创建该目录：

```
mkdir -p /data/db
```

运行可执行文件 mongod：

```
<path to binary>/mongod
```

停止运行：

```
# ./mongod --shutdown
```

创建服务：

```
# mkdir ../log/
# ./mongod --fork --logpath ../log/mongod.log
```

输出如下：

```
about to fork child process, waiting until server is ready for connections.
forked process: 30183
child process started successfully, parent exiting
```

默认路径在 /data/db：

```
# ls /data/db
```

在配置文件 mongod.conf 中定义绑定的 IP 地址，让 mongod 能够接收外部连接：

```
net:
   bindIp: 0.0.0.0
   port: 27017
```

使用配置文件：

```
# ./mongod -f /etc/mongod.conf
```

可以通过命令 mongoimport 把数据导入到 local 数据库：

```
# ./mongoimport -d local -c users --file ./sample_2.json --type json
```

也可以用 mongoexport 把数据导出。

首先查看帮助信息：

```
#./mongoexport --help
```

导出数据：

```
# ./mongoexport -d local -c ship -o ship.json
```

MongoDB 驱动程序允许用户使用不同的编程语言处理 MongoDB。PyMongo 是用于 Python 应用程序的官方 MongoDB 驱动程序。如下命令安装驱动程序：

```
>pip install pymongo
```

要连接到数据库，我们使用 MongoClient 类来访问 mongodb 实例，并通过它选择我们想要使用的数据库。这个类的无参数构造函数连接到端口 27017 上的示例：

```
from pymongo import MongoClient

# creation of MongoClient
client = MongoClient()
```

或者接受一个连接字符串：

```
client = MongoClient("mongodb://localhost:27017/")
```

插入一条数据：

```
# importing module
from pymongo import MongoClient

# creation of MongoClient
client = MongoClient()

# Connect with the portnumber and host
client = MongoClient()

# Access database
mydatabase = client['test']

print(mydatabase)

mycollection = mydatabase["news"]

news_rec1 = {
            "url":"https://news.sina.com.cn/o/2017-10-09/doc-ifymrqmq1572406.shtml",
            "title":"title",
            "body":"body"
            }

news_id1 = mycollection.insert_one(news_rec1)

print("Data inserted with ids",news_id1)
```

```
cursor = mycollection.find()
for record in cursor:
    print(record)

client.close()
```

爬虫抓进来的数据将会积压得越来越多，可以清空 MongoDB 中 n 天之前的数据，为了免去自己写个脚本去清除这些麻烦的数据，可以使用 ttlindex 来让这些数据保存 n 天，也就是 n 天之后自动删除。

TTL（Time To Live）索引是特定的单字段索引，MongoDB 可以在特定时间或特定时间段后自动从集合中删除文档。数据到期对于某些类型的信息（如机器生成的事件数据、日志和会话信息）很有用，因为这样的信息只需要在数据库中持续有限的时间。

要创建 TTL 索引，可以在其值为日期或包含日期值的数组的字段上使用 db.collection.createIndex() 方法，再加上 expireAfterSeconds 选项，例如，要在 eventlog 集合的 lastModifiedDate 字段上创建 TTL 索引，可以在 mongo shell 中使用以下操作：

```
db.eventlog.createIndex( { "lastModifiedDate": 1 }, { expireAfterSeconds: 3600 } )
```

3.16.4 存入 Elasticsearch 搜索引擎

可以通过客户端和 Elasticsearch 搜索引擎的服务器端打交道。elasticsearch-py(https://github.com/elastic/elasticsearch-py) 是 Elasticsearch 的官方底层 Python 客户端。

用 pip 安装 elasticsearch 模块：

```
pip install elasticsearch
```

索引和查询的示例：

```
from elasticsearch import Elasticsearch

es = Elasticsearch("http://localhost:9200")

e1={
"url":"https://news.sina.com.cn/o/2017-10-09/doc-ifymrqmq1572406.shtml",
"title": "title",
"body": "body"
}

res = es.index(index='news',id=1,document=e1)
print(res)
```

```
#res = es.search(index='news',body={'query':{'match_all':{}}})
res = es.search(index='news',query={'match_all':{}})
print("Got %d Hits:" % res['hits']['total']['value'])
print(res['hits']['hits'])
```

3.17 本章小结

Selenium 最初用于 Web 应用程序的自动化测试。本章介绍了使用 Selenium 抓取动态网页，重点熟悉了抓取网页的方法及通过提高 URL 的判重速度来提高抓取效率。

第 4 章 从互联网提取信息

文档用网络爬虫抓下来或者收集过来以后，需要转换成文字串才能索引入库。从 HTML 提取有效的文本，经常碰到两种类型的问题：一种是针对特定的网页特征提取结构化信息，还有一种就是通用的网页去噪。为了解决结构化信息提取的问题，本节介绍如何使用开源项目解析网页，以及如何把网页解析成 DOM 树，同时结合具体的例子介绍从 Web 网页提取文本的基本过程。

4.1 识别网页的编码

在实现从 Web 网页提取文本之前，首先要识别网页的编码，有时候还需要进一步识别网页所使用的语言，因为同一种编码可能对应多种语言，例如 UTF-8 编码可能对应英文或中文等任何语言。

识别编码的三个来源：Web 服务器返回的头信息；网页的 Meta 标签；返回流的二进制格式。

Web 服务器返回响应的头信息中可能包括了编码说明，例如：

```
HTTP/1.1 200 OK
Server: Apache-Coyote/1.1
Content-Type: text/html;charset=UTF-8
Content-Length: 5367
Date: Fri, 08 Apr 2011 11:08:24 GMT
Connection: close
```

网页的 Meta 标签，例如：

```
<meta http-equiv="Content-Type" content="text/html; charset=utf-8">
```

4.1.1 二进制流的编码

面前的一段文字，往往不用说明就能估计出是用什么文字写的。出现在不同语言中的字符有重叠，例如，可能会有出现在中文文档中的英文单词，所以要用统计的方法猜测二进制流的编码。

网页比单纯的文本要复杂。网页文本中有可能包含和编码无关的、额外的噪声数据，比如 HTML 的标记、空格和其他的格式/控制字符。

chardet（https://github.com/chardet/chardet）根据每种语言中的最常见的字符来估计二进制流是哪种编码。chardet 可以检测的字符编码有：中文、日文、韩文、西里尔文（Cyrillic）、希腊文、希伯来文。

使用 chardet 识别二进制编码的代码如下：

```
import chardet

str1 = ' 大家好，我是黄同学 '.encode('gbk')
print(chardet.detect(str1))  # 输出 {'encoding': 'GB2312', 'confidence': 0.99, 'language': 'Chinese'}
print(chardet.detect(str1)["encoding"])  # 输出 GB2312
```

借助 chardet 识别网页编码的代码如下：

```
import chardet
import requests

headers = {'User-Agent':'Mozilla/5.0 (Windows NT 6.1; WOW64) AppleWebKit/537.36 (KHTML, like Gecko) Chrome/55.0.2883.87 Safari/537.36'}
response = requests.get('https://www.baidu.com',headers=headers)

response.encoding = chardet.detect(response.content)['encoding']
print(response.encoding)
print(response.text)
```

response.apparent_encoding 就是使用了 chardet 模块从网页的内容中判断网页编码。

4.1.2 识别编码的整体流程

识别编码的整体流程如下：

（1）从 Web 服务器返回的 content type 头信息中提取编码，如果是 GB2312 类型的编码要当成 GBK 处理。

（2）从网页的 Meta 标签中识别字符编码，如果和 content type 中的编码不一致，以 Meta 中声明的编码为准。

（3）如果仍然无法确定网页所使用的字符集，需要从返回流的二进制格式判断。

（4）确定网页所使用的语言，往往采用统计的方法来估计网页的语言。

Requests 库的 utils.py 里的 get_encodings_from_content() 函数已经实现了从网页源代码

中获取 Meta 标签中的字符集。

```
import requests
url = "http://roll.news.sina.com.cn/news/gjxw/hqqw/index_3.shtml"
req = requests.get(url)
print(requests.utils.get_encodings_from_content(req.text))
```

完整的识别网页编码的代码如下：

```
import requests
url = "http://roll.news.sina.com.cn/news/gjxw/hqqw/index_3.shtml"
req = requests.get(url)

meta_charset = requests.utils.get_encodings_from_content(req.text)
if len(meta_charset)==0:
    req.encoding = req.apparent_encoding
else:
    req.encoding = meta_charset[0]
print(req.text)
```

4.2 正则表达式

可以用正则表达式提取字符串中的 Email 地址。可以在网站 http://regexpal.com/ 在线测试正则表达式，输入正则表达式和要匹配的文本，返回匹配出来的位置。

可以用正则表达式验证某个字符串是否符合指定的模式，或者说，是否可以接收一个字符串，例如匹配电话号码：

```
import re

pattern = "\\d\\d\\d([-\\s])?\\d\\d\\d\\d\\d\\d\\d\\d"
tel = "010-81727660"
result = re.fullmatch(pattern, tel)
print(result)
```

可以用 \d{3} 代替 \d\d\d，表示数字重复匹配三次，\d{8} 表示数字重复匹配八次。

```
pattern = "\\d{3}([-\\s])?\\d{8}"
tel = "010-81727660"
result = re.fullmatch(pattern, tel)
print(result)
```

检查 Email 的格式的语句：

```
mailTo = "abc@sina.com.cn"
```

```
pattern = r'([A-Za-z0-9]+[.-_])*[A-Za-z0-9]+@[A-Za-z0-9-]+(\.[A-Z|a-z]{2,})+'
result = re.fullmatch(pattern, mailTo)
print(result)
```

提取网页中的邮件地址:

```
target_string = "联系邮件: luogang@gmail.com"

pattern = r'([A-Za-z0-9]+[.-_])*[A-Za-z0-9]+@[A-Za-z0-9-]+(\.[A-Z|a-z]{2,})+'
# finditer() with regex pattern and target string
result = re.finditer(pattern, target_string)

# print all match object
for match_obj in result:
    # print each re.Match object
    print(match_obj)

    # extract each matching email
    print(match_obj.group())
```

提取电话号码:

```
htmlPage = "联系电话: 010-81727660"
pattern = "\\d{3}([-\\s])?\\d{8}"
result = re.finditer(pattern, htmlPage)

# print all match object
for match_obj in result:
    # print each re.Match object
    print(match_obj)

    # extract each matching tel
    print(match_obj.group())
```

4.3 结构化信息的提取

本节介绍从文本中提取结构化信息。介绍 Python 中的 XML 接口,然后介绍能够用 XML 接口从网页提取信息的项目 lxml。

4.3.1 解析 JSON

JSON(JavaScript Object Notation)是一种轻量级的数据交换格式。JSON 模块是一

个 Python 的 JSON 框架。json.loads() 方法可用于解析有效的 JSON 字符串并将其转换为 Python 字典，它主要用于将由 JSON 数据组成的原生字符串、字节或字节数组反序列化为 Python 字典，示例代码如下：

```
import json

# JSON string:
# Multi-line string
x = """{
    "Name": "Jennifer Smith",
    "Contact Number": 7867567898,
    "Email": "jen123@gmail.com",
    "Hobbies":["Reading", "Sketching", "Horse Riding"]
    }"""

# parse x:
y = json.loads(x)

# the result is a Python dictionary:
print(y)
```

4.3.2 解析 XML

可以使用 BeautifulSoup 解析 XML 文档，例如，获取网站 https://www.runoob.com/try/xml/books.xml 中 lang="en" 的 title 节点中的文字和 price 节点中的文字。

该网站的 XML 文档如下：

```
<?xml version="1.0" encoding="ISO-8859-1"?>
<bookstore>
<book category="COOKING">
<title lang="en">Everyday Italian</title>
<author>Giada De Laurentiis</author>
<year>2005</year>
<price>30.00</price>
</book>
<book category="CHILDREN">
<title lang="en">Harry Potter</title>
<author>J K. Rowling</author>
<year>2005</year>
<price>29.99</price>
</book>
<book category="WEB">
<title lang="en">XQuery Kick Start</title>
```

```
<author>James McGovern</author>
<author>Per Bothner</author>
<author>Kurt Cagle</author>
<author>James Linn</author>
<author>Vaidyanathan Nagarajan</author>
<year>2003</year>
<price>49.99</price>
</book>
<book category="WEB">
<title lang="en">Learning XML</title>
<author>Erik T. Ray</author>
<year>2003</year>
<price>39.95</price>
</book>
</bookstore>
```

实现解析的源代码如下：

```
import requests
from bs4 import BeautifulSoup

URL = "https://www.runoob.com/try/xml/books.xml"
r = requests.get(URL)

soup = BeautifulSoup(r.content,
                    'html5lib')  # If this line causes an error, run 'pip install html5lib' or install html5lib

nodes = soup.findAll('title',attrs = {'lang':'en'})

for row in nodes:
    p = row.find_parent()
    print(p.find('title').text)
    print(p.find('price').text)
```

4.3.3 XML 接口

一个 XML 文档可以看成由结点构成的树，假设有如下的 XML 文件。

```
<Names>
    <Name>
        <FirstName>John</FirstName>
        <LastName>Smith</LastName>
    </Name>
    <Name>
```

```
            <FirstName>James</FirstName>
            <LastName>White</LastName>
        </Name>
</Names>
```

上面的 XML 文档可以表示成如图 4-1 所示的 DOM 树：

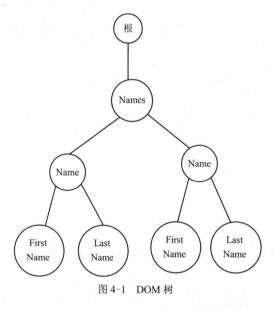

图 4-1　DOM 树

要找到一个 XML 文件中的节点，可以使用 XPath 表达式。使用 XPath 表达式 /Names/Name 可以得到所有 <Name> 节点。第一个斜杠表示该 <Names> 节点必须是一个根节点。xpath() 方法返回节点列表，这个列表包含一些 <Name> 节点。

lxml 是一个用 Python 编写的库，可以快速灵活地处理 XML 并支持 XPath。

为了将其安装在 Python 中，只需在控制台中使用以下命令。

```
pip install lxml
```

导入 lxml.etree 的常用方法如下：

```
>>> from lxml import etree
```

提取 <Name> 节点的代码如下：

```
x = """<Names>
    <Name>
        <FirstName>John</FirstName>
        <LastName>Smith</LastName>
    </Name>
    <Name>
        <FirstName>James</FirstName>
```

```
            <LastName>White</LastName>
        </Name>
</Names>
"""
root = etree.fromstring(x)

names = root.xpath('/Names/Name')
```

4.3.4　lxml 处理网页

HTML 本身其实只是一个 HTML 标记的字符串而已，因此一般说到要解析 HTML，第一个会想到的大概就是字符串查找，自己针对 HTML 的结构写一个模式，然后用 str.find() 方法来找，例如要找一个 td 标签：

```
pattern = "<td id='stockPrice'>"
html.find(pattern)
```

不过传统的字符串查找效率太差，也没有一个规则性，因而发展出了正则表达式技术，例如下列这样的语法：

```
" 小计: <em>(?<price>[\\d]*)</em>";
```

但正则表达式的可读性不好。

由于 HTML 标记的嵌套特性，可以采用 XPath 导航 HTML 文件的内部结构。

使用 lxml.html.fromstring() 方法加载网页。

```
import requests
import lxml.html

url = "https://www.imdb.com/search/title/?groups=top_1000&sort=user_rating,desc&count=200"
response = requests.get(url)
content = response.content
html = lxml.html.fromstring(content)
```

lxml 把网页转换成由节点组成的树。HTML 文档是 lxml.html.HtmlElement 类型。

4.3.5　使用 XPath 提取信息

lxml 可以把 HTML 文档转换成 etree 树。XPath 使用路径表达式来选取 etree 中的节点或者节点集。DOM 树可以用 XPath 来导航。

如果路径以单斜线 / 开始，那么该路径就表示到一个元素的绝对路径。例如，可以把正文的内容节点用 XPath 表示出来：

/HTML[1]/BODY[1]/DIV[1]/TABLE[1]/TBODY[1]/TR[1]/TD[1]/TABLE[1]/TBODY[1]/TR[1]/TD[1]

但是这种 XPath 的绝对路径的表示方式当 etree 树的节点有删除或修改后就失效了。还可以用 XPath 中的相对路径来选择节点，例如选择网页中所有链接的 XPath 是 "//a"，选择一个标题 DIV 可能是 "//div[@class='title']"。如果路径以双斜线 // 开头，则表示相对路径，选择文档中所有满足双斜线 // 之后规则的元素，例如要提取豆瓣电影的"一周口碑榜"：

```
import requests
from lxml import etree

url = 'http://movie.douban.com'
headers = {'User-agent': "Mozilla/5.0 (Windows NT 6.1) AppleWebKit/537.36 (KHTML, like Gecko) \
            Chrome/55.0.2883.75 Safari/537.36"}
response = requests.get(url, headers=headers)

with response:
    if response.status_code == 200:
        text = response.text
        html = etree.HTML(text)
        print(html.tag)

        titles = html.xpath('//div[@class="billboard-bd"]//a/text()')
        for title in titles:
            print(title)

        print("*********************")
```

4.3.6 在 Chrome 浏览器中查找 Selenium WebDriver 的 XPath

在 Chrome 浏览器中我们有两个用于查看网页中元素的 XPath 的选项，要么右键单击网页并单击检查选项，要么按 F12 键，之后，将能够看到任何 Web 元素的详细信息。

让我们首先打开百度首页，然后使用检查方式，右键单击输入框，然后您可以在右边面板看到元素和不同的选项来查看 XPath。

按 "Ctrl+F" 键，将在底部打开另一个小条，在这里可以按字符串、选择器或 XPath 查找元素。编写 XPath "//input[@id="kw"]"。

根据 XPath 查找元素的 Python 代码如下：

```
from selenium import webdriver
from selenium.webdriver.chrome.service import Service

from selenium.webdriver.common.by import By

from selenium.webdriver.support.wait import WebDriverWait

s = Service(r"D:/soft/chromedriver_win32/chromedriver.exe")
driver = webdriver.Chrome(service=s)

# Load the HTML page
driver.get("https://www.baidu.com")

driver.find_element(By.XPATH,"//input[@id=\"kw\"]").send_keys("test")
driver.find_element(By.ID, "su").click()
```

4.3.7 CSS 选择器

lxml 提供了 CSS 选择器来查询网页元素，使用 CSS 选择器的示例代码如下：

```
import requests
import lxml.html
from lxml.etree import tostring as htmlstring

url = "https://www.imdb.com/search/title/?groups=top_1000&sort=user_rating,desc&count=200"
response = requests.get(url)
content = response.content
html = lxml.html.fromstring(content)

print(htmlstring(html.cssselect('title')[0]))
```

层叠选择器目前的实现是从右至左解析的，这对于扫描少量无规律元素有利。但事实上一般在使用中，都是全文档搜索，这个时候从左至右就有很大的性能优势，尤其是层叠多次，或者左侧层叠条件苛刻的层叠选择器。

4.3.8 使用 Parsel

Parsel 是一个使用 XPath 和 CSS 选择器从 HTML 和 XML 中提取数据的库。

首先安装 Parsel 库：

```
pip install parsel
```

为了使用 Parsel 库，首先为要分析的 HTML 或 XML 文本创建 Selector 对象：

```
>>> from parsel import Selector
>>> text = "<html><body><h1>Hello, Parsel!</h1></body></html>"
>>> selector = Selector(text=text)
```

然后使用 CSS 或 XPath 表达式选择元素：

```
>>> selector.css('h1')
[<Selector xpath='descendant-or-self::h1' data='<h1>Hello, Parsel!</h1>'>]
>>> selector.xpath('//h1')    # the same, but now with XPath
[<Selector xpath='//h1' data='<h1>Hello, Parsel!</h1>'>]
```

并从这些元素中提取数据：

```
>>> selector.css('h1::text').get()
'Hello, Parsel!'
>>> selector.xpath('//h1/text()').getall()
['Hello, Parsel!']
```

抓取 https://quotes.toscrape.com/ 并输出标题的代码如下。

```
import parsel
import requests

URL = "https://quotes.toscrape.com/"

response = requests.get(URL).text

selector = parsel.Selector(text=response)

print(selector.css('title::text').get())
```

Parsel 库包括名为 get() 和 getall() 的函数，用于显示 title 标签的内容，例如爬取当当网的数据，搜索 Python 关键字的第一页的所有书籍信息：

```
import re
import parsel
import requests
url = 'https://search.dangdang.com/?key=python'
headersvalue = {'User-Agent': 'Mozilla/5.0 (Windows NT 10.0; Win64; x64) AppleWebKit/537.36 (KHTML, like Gecko) Chrome/107.0.0.0 Safari/537.36 Edg/107.0.1418.56'}
r = requests.get(url=url, headers=headersvalue)
selectors = parsel.Selector(r.text)
lis = selectors.css('ul.bigimg li ')
for li in lis:
```

```
            list1 = li.css(" a::attr(title)").get()
            list2 = li.css(".price span.search_now_price::text").get()
            list3 = li.css(".price span.search_pre_price::text").get()
            list4 = li.css(".search_book_author a::attr(title)").get()
            list5 = li.css('.search_book_author span:nth-child(2)::text ').get()
            list5 = re.sub("/", "", list5)
            list6 = li.css(".search_book_author span:nth-child(3)  a::attr(title)").get()
            print("书名:" + list1 + "|" + "现价:" + list2 + "|" + "原价:" + list3 + "|"
+ "作者:" + list4 + "|" + "出版时间:" + list5 + "|" + "出版社:", list6)
```

4.3.9 提取文本

在网页中不能使用小于号＜和大于号＞，这是因为浏览器会误认为它们是标签。小于号和大于号是保留字符。如果希望正确地显示保留字符，必须用字符实体（Character Entities）转义。

字符实体用 & 开始，以 ; 结束。可以用名字表示一个字符实体，例如，小于号叫作 lt，在网页中显示小于号，可以这样写：<。

也可以用编号表示一个字符实体。小于号的编号是 60，所以也可以用 < 显示小于号。

表示一个字符实体的格式是：

```
&entity_name;
```

或者

```
&#entity_number;
```

html 库包含的函数 html.unescape() 会把网页中的字符实体替换成原来的字符，例如：

```
print(html.unescape('&lt;dfdf&gt;'))  #把字符串 "&lt;dfdf&gt;" 转变成 "<dfdf>"
```

4.3.10 网页正文提取

网页有目录导航式页面（List Page）和详细页面（Detail Page）等。详细页面需要抽取的正文信息包括标题和内容等。

详细页面的特征有：

（1）文字较多，而且这些文字一般不在超级链接上。

（2）一般都有明显的文本段落，相应的标点符号也较多。

（3）URL 较长。在一般的 Web 网站链接导航树上，主题型网页主要分布于底层，多为叶节点。对于同一网站而言，主题型网页的 URL 相对较长。URL 体现了网站内容管理的层次，对于大型网站而言，URL 往往非常有规律。

（4）链接较少。主题型网页的主体在于"文字"，相对于导航型网页，其链接数较少。

详细页面中网页噪音的特征：

（1）多以链接的形式出现，链接到别的相关页面。

（2）有很多锚文本，但标点符号较少，锚文本往往是对其他链接页面的说明。

（3）有许多常见的噪音文本，如版权声明等，在视觉上，多出现于网页的边缘。

Goose3 库（https://github.com/goose3/goose3）是 Python 实现的 Web 页面内容提取工具库。

Goose 将尝试提取以下信息：

（1）一篇文章的主要内容；

（2）文章的主要图片；

（3）文章中嵌入的任何 YouTube/Vimeo 电影；

（4）元描述；

（5）元标签。

提取的代码如下：

```
from goose3 import Goose

url = 'http://edition.cnn.com/2012/02/22/world/europe/uk-occupy-london/index.html?hpt=ieu_c2'

g = Goose()

article = g.extract(url=url)

print(article.title)
print(article.cleaned_text)
print(article.meta_description)
```

有两种方法可以将配置传递给 Goose。第一个方法是向 Goose 传递一个 Configuration() 对象；第二个方法是传递一个配置字典。

例如，如果想更改 Goose 使用的 userAgent，只需通过如下代码配置：

```
>>> g = Goose({'browser_user_agent': 'Mozilla'})
```

Goose 现在可以与 lxml html 解析器或 lxml soup 解析器一起使用，默认情况下使用 html 解析器。如果想使用 soup 解析器，请在配置 dict 中传递它：

```
>>> g = Goose({'browser_user_agent': 'Mozilla', 'parser_class':'soup'})
```

4.4 从文件提取信息

textract 库（https://github.com/deanmalmgren/textract）提供了用于从任何类型的文件中提取内容的单一接口。

安装 textract 包：

```
>pip install textract
```

实现内容提取的 Python 代码如下：

```
import textract
text = textract.process("d:/test/test.docx")

print(text)
```

还可以将关键字参数传递给 textract.process() 函数，例如，使用特定方法解析 PDF，如下所示：

```
import textract
text = textract.process('path/to/a.pdf', method='pdfminer')
```

或指定特定的输出编码：

```
import textract
text = textract.process('path/to/file.extension', encoding='ascii')
```

当文件名没有扩展名时，可以将文件的扩展名指定为 textract.process 的参数，如下所示：

```
import textract
text = textract.process('path/to/file', extension='docx')
```

4.5 本章小结

本章介绍了使用 chardet 库识别网页编码，chardet 的算法来源于 UniversalCharDet。
同时介绍了使用正则表达式和网页解析器从网页源代码提取信息。
可以使用 textract 从文档提取信息。如果要专门提取 PDF 文件，还可以使用 pypdf 库。

第 5 章 使用 Scrapy 开发爬虫

本章介绍使用 Scrapy 框架开发爬虫。最后介绍开发 Scrapy 的 Twisted 框架。

5.1 一个示例爬虫的演练

为了展示 Scrapy 带来了什么，我们将引导您通过一个 Scrapy 爬虫的示例使用最简单的方式运行爬虫。

这是一个爬虫的代码，它从网站 https://quotes.toscrape.com 上抓取著名的名言：

```python
import scrapy

class QuotesSpider(scrapy.Spider):
    name = 'quotes'
    start_urls = [
        'https://quotes.toscrape.com/tag/humor/',
    ]

    def parse(self, response):
        for quote in response.css('div.quote'):
            yield {
                'author': quote.xpath('span/small/text()').get(),
                'text': quote.css('span.text::text').get(),
            }

        next_page = response.css('li.next > a::attr("href")').get()
        if next_page is not None:
            yield response.follow(next_page, self.parse)
```

将它放在一个文本文件中，将其命名为 quotes_spider.py 并使用 runspider 命令运行爬虫：

```
scrapy runspider quotes_spider.py -o quotes.jl
```

完成后，将在 quote.jl 文件中获得 JSON Lines 格式的名言列表，其中包含文本和作者，如下所示：

```
    {"author": "Jane Austen", "text": "\u201cThe person, be it gentleman or lady, who
has not pleasure in a good novel, must be intolerably stupid.\u201d"}
    {"author": "Steve Martin", "text": "\u201cA day without sunshine is like, you
know, night.\u201d"}
    {"author": "Garrison Keillor", "text": "\u201cAnyone who thinks sitting in church can
make you a Christian must also think that sitting in a garage can make you a car.\u201d"}
```

当运行命令 scrapy runspider quotes_spider.py 时，Scrapy 在 quotes_spider.py 中查找 Spider 定义并通过其爬虫引擎运行它。

爬网首先向 start_urls 属性中定义的 URL 发出请求（在这种情况下，只有幽默类别中的名言 URL）并调用默认回调方法 parse()，将响应对象作为参数传递。在解析回调中，使用 CSS 选择器循环名言元素，生成一个带有提取的名言文本和作者的 Python 字典，寻找到下一页的链接，并使用与回调相同的解析方法安排另一个请求。

在这里，你会注意到 Scrapy 的主要优点之一：请求是异步调度和处理的。这意味着 Scrapy 不需要等待请求完成和处理，它可以同时发送另一个请求或做其他事情，这也意味着即使某些请求失败或在处理它时发生错误，其他请求也可以继续进行。

虽然这使您能够进行非常快速的爬取（以容错方式同时发送多个并发请求），但 Scrapy 还可以通过一些设置让您控制爬取的礼貌。你可以执行以下操作：比如在每个请求之间设置下载延迟，限制每个域或每个 IP 的并发请求数量，甚至使用自动限制扩展来尝试自动计算这些请求。

5.2 Scrapy Playwright 指南：渲染和抓取动态 JS 网站

Playwright.js 由微软于 2020 年发布，由于其跨浏览器支持（可以驱动 Chromium、WebKit 和 Firefox 浏览器，而 Puppeter 只能驱动 Chromium）以及相对 Puppeter 更好的开发人员体验改进，它正迅速成为浏览器自动化和网络抓取最受欢迎的无头浏览器库。因此，很高兴看到许多 Scrapy 的核心维护人员为 Scrapy 开发了一个 Playwright 集成：scrapy-playwright(https://github.com/scrapy-plugins/scrapy-playwright)。

Scrapy Playwright 是您可以与 Scrapy 一起使用的最佳无头浏览器选项之一。

如果你想跟随一个已经设置好并准备就绪的项目，你可以克隆我们专门用于本节的 Scrapy 项目。

一旦从 GitHub 存储库（https://github.com/python-scrapy-playbook/quotes-js-project）下载了代码，您只需复制/粘贴在下面使用的代码片段，就可以看到代码在计算机上正常工作。

在 Scrapy 项目中安装 scrapy-playwright 是非常简单的。

首先，需要安装 scrapy-playwright 本身：

```
pip install scrapy-playwright
```

然后，如果还没有安装 Playwright 本身，则需要在命令行中使用以下命令进行安装：

```
playwright install
```

接下来，需要更新 Scrapy 项目设置，以激活项目中的 scrapy-playwright：

```
# settings.py

DOWNLOAD_HANDLERS = {
    "http": "scrapy_playwright.handler.ScrapyPlaywrightDownloadHandler",
    "https": "scrapy_playwright.handler.ScrapyPlaywrightDownloadHandler",
}

TWISTED_REACTOR = "twisted.internet.asyncioreactor.AsyncioSelectorReactor"
```

ScrapyPlaywrightDownloadHandler 类继承自 Scrapy 的默认 http/https 处理程序。因此，除非在 Scrapy 请求中明确激活 Scrapy，否则这些请求将由常规的 Scrapy 下载处理程序处理。

现在，将 scrapy-playwright 集成到 Scrapy 爬虫中，这样所有的请求都将被 JS 渲染。

要通过 scrapy-playwright 路由请求，只需要在请求元字典中通过设置 meta={'playwright': True} 来启用它。

```
# spiders/quotes.py

import scrapy
from quotes_js_scraper.items import QuoteItem

class QuotesSpider(scrapy.Spider):
    name = 'quotes'

    def start_requests(self):
        url = "https://quotes.toscrape.com/js/"
        yield scrapy.Request(url, meta={'playwright': True})

    def parse(self, response):
        for quote in response.css('div.quote'):
            quote_item = QuoteItem()
            quote_item['text'] = quote.css('span.text::text').get()
            quote_item['author'] = quote.css('small.author::text').get()
            quote_item['tags'] = quote.css('div.tags a.tag::text').getall()
            yield quote_item
```

response 现在将包含浏览器看到的渲染页面。然而，有时 Playwright 会在渲染整个页面之前结束渲染，可以使用 Playwright PageMethods 来解决这个问题。

要使用 scrapy-playwright 与页面交互，我们需要使用 PageMethod 类。PageMethod 允许我们在页面上做很多不同的事情，包括：

（1）等待元素加载后再返回响应；
（2）滚动页面；
（3）单击页面元素；
（4）对页面进行屏幕截图；
（5）创建页面的 PDF。

首先，要在爬虫中使用 PageMethod 功能，需要将 playwright_include_page 设置为 True，这样就可以访问 Playwright Page 对象，还可以将任何回调定义为协程函数，以等待提供的 Page 对象。

```python
# spiders/quotes.py

import scrapy
from quotes_js_scraper.items import QuoteItem


class QuotesSpider(scrapy.Spider):
    name = 'quotes'

    def start_requests(self):
        url = 'https://quotes.toscrape.com/js/'
        yield scrapy.Request(url, meta=dict(
            playwright = True,
            playwright_include_page = True,
        ))

    async def parse(self, response):
        ...
```

注意：当设置 'playwright_include_page': True 时，还建议您设置一个请求 errback，以确保即使请求失败，页面也会关闭（如果 playwright_include_page=False 或 unset，则页面在遇到异常时会自动关闭）。

```python
# spiders/quotes.py

import scrapy
from quotes_js_scraper.items import QuoteItem
```

```python
class QuotesSpider(scrapy.Spider):
    name = 'quotes'

    def start_requests(self):
        url = 'https://quotes.toscrape.com/js/'
        yield scrapy.Request(url, meta=dict(
            playwright = True,
            playwright_include_page = True,
            errback=self.errback,
        ))

    async def parse(self, response):
        page = response.meta["playwright_page"]
        await page.close()

        for quote in response.css('div.quote'):
            quote_item = QuoteItem()
            quote_item['text'] = quote.css('span.text::text').get()
            quote_item['author'] = quote.css('small.author::text').get()
            quote_item['tags'] = quote.css('div.tags a.tag::text').getall()
            yield quote_item

    async def errback(self, failure):
        page = failure.request.meta["playwright_page"]
        await page.close()
```

要在停止 javascript 渲染并向爬虫返回响应之前等待特定的页面元素，我们只需要在 Playwright 设置中的 playwright_page_methods 键中添加一个 PageMethod，并定义一个 wait_for_selector。

现在，当运行爬虫时，scrapy-playwright 将渲染页面，直到页面上出现一个带有类 quote 的 div。

```python
# spiders/quotes.py

import scrapy
from quotes_js_scraper.items import QuoteItem
from scrapy_playwright.page import PageMethod

class QuotesSpider(scrapy.Spider):
    name = 'quotes'

    def start_requests(self):
        url = "https://quotes.toscrape.com/js/"
        yield scrapy.Request(url, meta=dict(
```

```python
                playwright = True,
                playwright_include_page = True,
                playwright_page_methods =[PageMethod('wait_for_selector', 'div.quote')],
            errback=self.errback,
            ))

    async def parse(self, response):
        page = response.meta["playwright_page"]
        await page.close()

        for quote in response.css('div.quote'):
            quote_item = QuoteItem()
            quote_item['text'] = quote.css('span.text::text').get()
            quote_item['author'] = quote.css('small.author::text').get()
            quote_item['tags'] = quote.css('div.tags a.tag::text').getall()
            yield quote_item

    async def errback(self, failure):
        page = failure.request.meta["playwright_page"]
        await page.close()
```

通常需要在 javascript 渲染的网站上抓取多个页面，可以通过检查页面上是否存在下一个页面链接来实现这一点，然后使用从页面中抓取的 URL 请求该页面。

```python
# spiders/quotes.py

import scrapy
from quotes_js_scraper.items import QuoteItem
from scrapy_playwright.page import PageMethod


class QuotesSpider(scrapy.Spider):
    name = 'quotes'

    def start_requests(self):
        url = "https://quotes.toscrape.com/js/"
        yield scrapy.Request(url, meta=dict(
            playwright = True,
            playwright_include_page = True,
            playwright_page_methods =[
                PageMethod('wait_for_selector', 'div.quote'),
            ],
        errback=self.errback,
            ))
```

```python
    async def parse(self, response):
        page = response.meta["playwright_page"]
        await page.close()

        for quote in response.css('div.quote'):
            quote_item = QuoteItem()
            quote_item['text'] = quote.css('span.text::text').get()
            quote_item['author'] = quote.css('small.author::text').get()
            quote_item['tags'] = quote.css('div.tags a.tag::text').getall()
            yield quote_item

        next_page = response.css('.next>a ::attr(href)').get()

        if next_page is not None:
            next_page_url = 'http://quotes.toscrape.com' + next_page
            yield scrapy.Request(next_page_url, meta=dict(
                playwright = True,
                playwright_include_page = True,
                playwright_page_methods =[
                    PageMethod('wait_for_selector', 'div.quote'),
                ],
                errback=self.errback,
            ))

    async def errback(self, failure):
        page = failure.request.meta["playwright_page"]
        await page.close()
```

当一个网站使用无限滚动加载数据时,也可以配置 scrapy-playwright 向下滚动页面。

在本例中,在向下滚动页面,直到到达第 10 个名言之前,Playwright 将等待 div.quote 出现。

```python
# spiders/quotes.py

import scrapy
from quotes_js_scraper.items import QuoteItem
from scrapy_playwright.page import PageMethod

class QuotesSpider(scrapy.Spider):
    name = 'quotes'

    def start_requests(self):
        url = "https://quotes.toscrape.com/scroll"
        yield scrapy.Request(url, meta=dict(
            playwright = True,
```

```python
                playwright_include_page = True,
                playwright_page_methods =[
            PageMethod("wait_for_selector", "div.quote"),
            PageMethod("evaluate", "window.scrollBy(0, document.body.scrollHeight)"),
            PageMethod("wait_for_selector", "div.quote:nth-child(11)"),  # 10 per page
            ],
            errback=self.errback,
            ))

    async def parse(self, response):
        page = response.meta["playwright_page"]
        await page.close()

        for quote in response.css('div.quote'):
            quote_item = QuoteItem()
            quote_item['text'] = quote.css('span.text::text').get()
            quote_item['author'] = quote.css('small.author::text').get()
            quote_item['tags'] = quote.css('div.tags a.tag::text').getall()
            yield quote_item

    async def errback(self, failure):
        page = failure.request.meta["playwright_page"]
        await page.close()
```

页面截图也很简单。在这里，当 Playwright 看到选择器 div.quote，然后它会截取页面的屏幕截图。

```python
# spiders/quotes.py

import scrapy
from quotes_js_scraper.items import QuoteItem
from scrapy_playwright.page import PageMethod

class QuotesSpider(scrapy.Spider):
    name = 'quotes'

    def start_requests(self):
        url = "https://quotes.toscrape.com/js/"
        yield scrapy.Request(url, meta=dict(
            playwright = True,
            playwright_include_page = True,
            playwright_page_methods =[
            PageMethod("wait_for_selector", "div.quote"),
            ],
            ))
```

```
async def parse(self, response):
    page = response.meta["playwright_page"]
    screenshot = await page.screenshot(path="example.png", full_page=True)
    # screenshot contains the image's bytes
    await page.close()
```

在 Scrapy Playwright 中，可以在 PLAYWRIGHT_LAUNCH_OPTIONS 设置中指定 proxy 键。在浏览器级别配置代理如下：

```
# spiders/quotes.py

from scrapy import Spider, Request

class ProxySpider(Spider):
    name = "proxy"
    custom_settings = {
        "PLAYWRIGHT_LAUNCH_OPTIONS": {
            "proxy": {
                "server": "http://myproxy.com:3128",
                "username": "user",
                "password": "pass",
            },
        }
    }

    def start_requests(self):
        yield Request("http://httpbin.org/get", meta={"playwright": True})

    def parse(self, response):
        print(response.text)
```

5.3 将抓取的数据保存到 SQLite 数据库

每个爬虫项目需要做的最常见的任务之一就是保存我们抓取的数据。在保存数据时，我们可以选择许多选项，但是，当您有一个小项目时，使用 SQLite 是最好的选择之一。

在本节中，我们将介绍如何使用 Scrapy 管道将数据保存到 SQLite 数据库。

项目管道是 Scrapy 处理爬虫抓取的数据的方式。

在爬虫抓取一个项目后，它被发送到项目管道，该管道通过一系列步骤对其进行处理，这些步骤可以配置为清理和处理抓取的数据，然后最终将其保存在某个地方。

可以使用项目管道进行如下操作：

（1）清理 HTML 数据；

（2）验证抓取的数据；

（3）检查和删除重复数据；

（4）将数据存储在数据库中。

这里将重点介绍使用 Item Pipelines 在 SQLite 数据库中存储数据。

创建一个名为 sqlite_demo 的 Scrapy 项目：

```
scrapy startproject sqlite_demo
```

打开 pipelines.py 文件并设置管道。打开 pipelines.py 文件时，默认文件应如下所示：

```
# pipelines.py

from itemadapter import ItemAdapter

class SqliteDemoPipeline:
    def process_item(self, item, spider):
        return item
```

生成爬虫：

```
scrapy genspider quotes toscrape.com
```

修改 items.py 内容如下：

```
from scrapy.item import Item, Field

class QuoteItem(Item):
    text = Field()
    tags = Field()
    author = Field()
```

修改 quotes.py 内容如下：

```
import scrapy
from sqlite_demo.items import QuoteItem

class QuotesSpider(scrapy.Spider):
    name = 'quotes'
    def start_requests(self):
        url = 'https://quotes.toscrape.com/'
        yield scrapy.Request(url, callback=self.parse)

    def parse(self, response):
```

```
            quote_item = QuoteItem()
            for quote in response.css('div.quote'):
                quote_item['text'] = quote.css('span.text::text').get()
                quote_item['author'] = quote.css('small.author::text').get()
                quote_item['tags'] = quote.css('div.tags a.tag::text').getall()
                yield quote_item
```

Python 附带 SQLite，因此无须安装任何东西即可在 Scrapy 项目中使用它。

首先，我们将 sqlite3 模块导入到 pipelines.py 文件中，并创建一个 __init__ 方法，我们将使用它来创建数据库和表。

```
# pipelines.py

import sqlite3

class SqliteDemoPipeline:

    def __init__(self):
        pass

    def process_item(self, item, spider):
        return item
```

在 __init__ 方法中，我们将配置管道以在每次管道被爬虫激活时执行以下操作：

（1）尝试连接到数据库 demo.db，但如果它不存在，则创建数据库。

（2）创建一个游标，我们将使用它在数据库中执行 SQL 命令。

（3）如果数据库中不存在名言表，则创建一个包括文本、标签和作者列的新表。

```
# pipelines.py

import sqlite3

class SqliteDemoPipeline:

    def __init__(self):

        # Create/Connect to database
        self.con = sqlite3.connect('demo.db')

        # Create cursor, used to execute commands
        self.cur = self.con.cursor()

        # Create quotes table if none exists
        self.cur.execute("""
```

```
        CREATE TABLE IF NOT EXISTS quotes(
            text TEXT,
            tags TEXT,
            author TEXT
        )
        """)

    def process_item(self, item, spider):
        return item
```

接下来，我们将使用 Scrapy 管道中的 process_item 事件将我们抓取的数据存储到 SQLite 数据库中。

process_item() 将在我们的爬虫每次抓取项目时激活，因此我们需要配置 process_item() 方法以将项目数据插入到数据库中。

```
# pipelines.py

import sqlite3

class SqliteDemoPipeline:

    def __init__(self):

        # Create/Connect to database
        self.con = sqlite3.connect('demo.db')

        # Create cursor, used to execute commands
        self.cur = self.con.cursor()

        # Create quotes table if none exists
        self.cur.execute("""
        CREATE TABLE IF NOT EXISTS quotes(
            text TEXT,
            tags TEXT,
            author TEXT
        )
        """)


    def process_item(self, item, spider):

        # Define insert statement
        self.cur.execute("""
            INSERT INTO quotes (text, tags, author) VALUES (?, ?, ?)
```

```
            """,
            (
                item['text'],
                str(item['tags']),
                item['author']
            ))

            # Execute insert of data into database
            self.con.commit()
            return item
```

在这里,我们首先定义了 SQL 插入语句并为其提供了数据(注意,这里对标签值进行了字符串化,因为它是一个数组),然后我们使用了 self.con.commit() 命令插入数据。

最后,要激活项目管道,需要将它包含在 settings.py 文件中:

```
# settings.py

ITEM_PIPELINES = {
   'sqlite_demo.pipelines.SqliteDemoPipeline': 300,
}
```

现在,当我们运行我们的名言爬虫时,SqliteDemoPipeline 会将所有抓取的项目存储在数据库中。

好的,现在我们有了一个项目管道,可以将所有抓取的项目保存到我们的 SQLite 数据库中。但是,如果我们只想保存以前没有抓过的新数据怎么办。

我们可以轻松地重新配置我们的管道来执行此操作,方法是让它在再次插入之前检查数据是否已经在数据库中。

为此,将在 pipelines.py 文件中创建一个名为 SqliteNoDuplicatesPipeline 的新管道,并更改 process_item() 方法,使其仅将新数据插入到数据库中。

它将首先在数据库中查找 item['text'],并且只有在它不存在时才会插入到新项目中。

```
# pipelines.py

import sqlite3

class SqliteNoDuplicatesPipeline:

    def __init__(self):

        # Create/Connect to database
        self.con = sqlite3.connect('demo.db')
```

```python
        # Create cursor, used to execute commands
        self.cur = self.con.cursor()

        # Create quotes table if none exists
        self.cur.execute("""
        CREATE TABLE IF NOT EXISTS quotes(
            text TEXT,
            tags TEXT,
            author TEXT
        )
        """)


    def process_item(self, item, spider):

        # Check to see if text is already in database
        self.cur.execute("select * from quotes where text = ?", (item['text'],))
        result = self.cur.fetchone()

        # If it is in DB, create log message
        if result:
            spider.logger.warn("Item already in database: %s" % item['text'])

        # If text isn't in the DB, insert data
        else:

            # Define insert statement
            self.cur.execute("""
                INSERT INTO quotes (text, tags, author) VALUES (?, ?, ?)
            """,
            (
                item['text'],
                str(item['tags']),
                item['author']
            ))

            # Execute insert of data into database
            self.con.commit()

        return item
```

要激活此管道，我们还需要更新 settings.py 以使用 SqliteNoDuplicatesPipeline 而不是之前的 SqliteDemoPipeline 管道：

```
# settings.py
```

```
ITEM_PIPELINES = {
    # 'sqlite_demo.pipelines.SqliteDemoPipeline': 300,
    'sqlite_demo.pipelines.SqliteNoDuplicatesPipeline': 300,
}
```

现在，当我们运行我们的名言爬虫时，管道将只存储尚未在数据库中的新数据。

5.4 将抓取的数据保存到 MySQL 数据库

如果您抓取网站，则需要将该数据保存在某处。MySQL 是一个不错的选择，它是目前最流行且易于使用的 SQL 数据库之一。

在本节中，我们将介绍如何使用 Scrapy 管道将数据保存到 MySQL 数据库。

为了与数据库交互，需要一个库来处理交互。为此将安装 mysql 和 mysql-connector-python 库。

```
pip install mysql mysql-connector-python
```

我们将使用 mysql 库与 MySQL 数据库进行交互。

创建一个名为 mysql_demo 的 Scrapy 项目：

```
scrapy startproject mysql_demo
```

生成爬虫：

```
scrapy genspider quotes toscrape.com
```

修改 items.py 内容如下：

```
from scrapy.item import Item, Field

class QuoteItem(Item):
    text = Field()
    tags = Field()
    author = Field()
```

修改 quotes.py 内容如下：

```
# spiders/quotes.py

import scrapy
from mysql_demo.items import QuoteItem
```

```python
class QuotesSpider(scrapy.Spider):
    name = 'quotes'

    def start_requests(self):
        url = 'https://quotes.toscrape.com/'
        yield scrapy.Request(url, callback=self.parse)

    def parse(self, response):
        quote_item = QuoteItem()
        for quote in response.css('div.quote'):
            quote_item['text'] = quote.css('span.text::text').get()
            quote_item['author'] = quote.css('small.author::text').get()
            quote_item['tags'] = quote.css('div.tags a.tag::text').getall()
            yield quote_item
```

下一步是需要打开 pipelines.py 文件并设置管道。

```python
# pipelines.py

from itemadapter import ItemAdapter

class MysqlDemoPipeline:
    def process_item(self, item, spider):
        return item
```

首先，我们将 mysql 模块导入到我们的 pipelines.py 文件中，并创建一个 __init__ 方法，我们将使用它来创建数据库和表。

```python
# pipelines.py

import mysql.connector

class MysqlDemoPipeline:

    def __init__(self):
        pass

    def process_item(self, item, spider):
        return item
```

在 __init__ 方法中，我们将配置管道以在每次管道被爬虫激活时执行以下操作：
（1）尝试连接到数据库 quotes，如果它不存在，则创建数据库。
（2）创建一个游标，将使用它在数据库中执行 SQL 命令。
（3）如果数据库中尚不存在 quotes 表，则创建一个包含列内容、标签和作者的新表。

```python
# pipelines.py
```

```python
import mysql.connector

class MysqlDemoPipeline:

    def __init__(self):
        self.conn = mysql.connector.connect(
            host = 'localhost',
            user = 'root',
            password = '******',
            database = 'quotes'
        )

        # Create cursor, used to execute commands
        self.cur = self.conn.cursor()

        # Create quotes table if none exists
        self.cur.execute("""
        CREATE TABLE IF NOT EXISTS quotes(
            id int NOT NULL auto_increment,
            content text,
            tags text,
            author VARCHAR(255),
            PRIMARY KEY (id)
        )
        """)

    def process_item(self, item, spider):
        return item
```

接下来,我们将使用 Scrapy 管道中的 process_item 事件将抓取的数据存储到 MySQL 数据库中。

process_item 将在爬虫每次抓取项目时激活,因此我们需要配置 process_item() 方法以将项目数据插入数据库中。

我们将在爬虫关闭时调用 close_spider() 方法来关闭游标和数据库的连接,以避免保持连接打开。

```python
# pipelines.py

import mysql.connector

class MysqlDemoPipeline:
```

```python
    def __init__(self):
        self.conn = mysql.connector.connect(
            host = 'localhost',
            user = 'root',
            password = '**********',
            database = 'quotes'
        )

        # Create cursor, used to execute commands
        self.cur = self.conn.cursor()

        # Create quotes table if none exists
        self.cur.execute("""
        CREATE TABLE IF NOT EXISTS quotes(
            id int NOT NULL auto_increment,
            content text,
            tags text,
            author VARCHAR(255),
            PRIMARY KEY (id)
        )
        """)


    def process_item(self, item, spider):

        # Define insert statement
        self.cur.execute(""" insert into quotes (content, tags, author) values (%s,%s,%s)""", (
            item["text"],
            str(item["tags"]),
            item["author"]
        ))

        # Execute insert of data into database
        self.conn.commit()


    def close_spider(self, spider):

        # Close cursor & connection to database
        self.cur.close()
        self.conn.close()
```

最后，要激活我们的项目管道，我们需要将它包含在 settings.py 文件中：

```
# settings.py
```

```
ITEM_PIPELINES = {
    'mysql_demo.pipelines.MysqlDemoPipeline': 300,
}
```

现在,当我们运行名言爬虫时,MysqlDemoPipeline 会将所有抓取的项目存储在数据库中。

现在我们有一个项目管道,可以将所有抓取的项目保存到 MySQL 数据库中。但是,如果我们只想保存以前没有抓取过的新数据怎么办呢?

我们可以通过检查项目是否已经在数据库中,然后再插入数据库来重新配置管道执行此操作。

为此,将在 pipelines.py 文件中创建一个名为 MySQLNoDuplicatesPipeline 的新管道,并更改 process_item() 方法,使其仅将新数据插入数据库中。

首先将在数据库中查找 item['text'],并且只有在它不存在时才会插入新项目。

```python
# pipelines.py

import mysql.connector

class MySQLNoDuplicatesPipeline:

    def __init__(self):
        self.conn = mysql.connector.connect(
            host = 'localhost',
            user = 'root',
            password = 'Eily1990',
            database = 'quotes'
        )

        # Create cursor, used to execute commands
        self.cur = self.conn.cursor()

        # Create quotes table if none exists
        self.cur.execute("""
        CREATE TABLE IF NOT EXISTS quotes(
            id int NOT NULL auto_increment,
            content text,
            tags text,
            author VARCHAR(255),
            PRIMARY KEY (id)
        )
        """)
```

```python
    def process_item(self, item, spider):

        # Check to see if text is already in database
        self.cur.execute("select * from quotes where content = %s", (item['text'],))
        result = self.cur.fetchone()

        # If it is in DB, create log message
        if result:
            spider.logger.warn("Item already in database: %s" % item['text'])

        # If text isn't in the DB, insert data
        else:

            # Define insert statement
            self.cur.execute(""" insert into quotes (content, tags, author) values (%s,%s,%s)""", (
                item["text"],
                str(item["tags"]),
                item["author"]
            ))

            # Execute insert of data into database
            self.connection.commit()
        return item


    def close_spider(self, spider):

        # Close cursor & connection to database
        self.cur.close()
        self.conn.close()
```

要激活这个管道,还需要更新 settings.py 以使用 MySQLNoDuplicatesPipeline 而不是之前的 MysqlDemoPipeline 管道:

```
# settings.py

ITEM_PIPELINES = {
#     'mysql_demo.pipelines.MysqlDemoPipeline': 300,
    'mysql_demo.pipelines.MySQLNoDuplicatesPipeline': 300,
```

```
}
```

现在,当我们运行名言爬虫时,管道将只存储尚未在数据库中的新数据。

5.5 将抓取的数据保存到 Postgres 数据库

在本节中,将介绍如何使用 Scrapy 管道将数据保存到 Postgres 数据库。

为了开始,首先需要设置一个 Postgres 数据库。可以使用以下下载(https://www.postgresql.org/download/)之一在本地计算机上设置一个数据库。

设置后,可以访问数据库的数据库连接的详细信息:

```
host="localhost",
database="my_database",
user="root",
password="123456"
```

现在集成保存数据到 Postgres 数据库中。

为了与数据库进行交互,需要一个库来处理交互,为此,将安装 psycopg2。

```
pip install psycopg2
```

将使用 psycopg2 与 Postgres 数据库进行交互。

下一步是打开 pipelines.py 文件并设置管道。打开 pipelines.py 文件时,默认文件应如下所示:

```
# pipelines.py

from itemadapter import ItemAdapter

class PostgresDemoPipeline:
    def process_item(self, item, spider):
        return item
```

现在将配置这个空管道来存储数据。

在本节中,创建了一个名为 postgres_demo 的 Scrapy 项目(因此默认管道是 PostgresDemoPipeline),并使用了这个爬虫:

```
# spiders/quotes.py

import scrapy
from postgres_demo.items import QuoteItem
```

```python
class QuotesSpider(scrapy.Spider):
    name = 'quotes'

    def start_requests(self):
        url = 'https://quotes.toscrape.com/'
        yield scrapy.Request(url, callback=self.parse)

    def parse(self, response):
        quote_item = QuoteItem()
        for quote in response.css('div.quote'):
            quote_item['text'] = quote.css('span.text::text').get()
            quote_item['author'] = quote.css('small.author::text').get()
            quote_item['tags'] = quote.css('div.tags a.tag::text').getall()
            yield quote_item
```

以及项目：

```python
# items.py

from scrapy.item import Item, Field

class QuoteItem(Item):
    text = Field()
    tags = Field()
    author = Field()
```

首先，将把 psycopg2 导入 pipelines.py 文件，并创建一个用于创建数据库和表的 __init__ 方法。

```python
# pipelines.py

import psycopg2

class PostgresDemoPipeline:

    def __init__(self):
        pass

    def process_item(self, item, spider):
        return item
```

在 __init__ 方法中，每当管道被爬虫激活时，将配置管道执行以下操作。

（1）尝试连接到数据库 quotes，但如果它不存在，则创建数据库。

（2）创建一个光标，用于在数据库中执行 SQL 命令。

（3）如果数据库中还不存在表 quotes，则创建一个包含列 content、tags 和 author 的新表。

```python
# pipelines.py

import psycopg2

class PostgresDemoPipeline:

    def __init__(self):
        # Connection Details
        hostname = 'localhost'
        username = 'postgres'
        password = '*******' # your password
        database = 'quotes'

        # Create/Connect to database
        self.connection = psycopg2.connect(host=hostname, user=username, password=password, dbname=database)

        # Create cursor, used to execute commands
        self.cur = self.connection.cursor()

        # Create quotes table if none exists
        self.cur.execute("""
        CREATE TABLE IF NOT EXISTS quotes(
            id serial PRIMARY KEY,
            content text,
            tags text,
            author VARCHAR(255)
        )
        """)

    def process_item(self, item, spider):
        return item
```

接下来，我们将使用 Scrapy 管道中的 process_item 事件来将我们抓取的数据存储到 Postgres 数据库中。

爬虫每次抓取一个项目，process_item 都会被激活，所以需要配置 process_item 方法来将项目数据插入到数据库中。

将在爬虫关闭时调用 close_spider() 方法来关闭与游标和数据库的连接，以避免保持连接打开。

```python
# pipelines.py
```

```python
import psycopg2

class PostgresDemoPipeline:

    def __init__(self):
        # Connection Details
        hostname = 'localhost'
        username = 'postgres'
        password = '******' # your password
        database = 'quotes'

        # Create/Connect to database
        self.connection = psycopg2.connect(host=hostname, user=username, password=password, dbname=database)

        # Create cursor, used to execute commands
        self.cur = self.connection.cursor()

        # Create quotes table if none exists
        self.cur.execute("""
        CREATE TABLE IF NOT EXISTS quotes(
            id serial PRIMARY KEY,
            content text,
            tags text,
            author VARCHAR(255)
        )
        """)

    def process_item(self, item, spider):

        # Define insert statement
        self.cur.execute(""" insert into quotes (content, tags, author) values (%s,%s,%s)""", (
            item["text"],
            str(item["tags"]),
            item["author"]
        ))

        # Execute insert of data into database
        self.connection.commit()
        return item

    def close_spider(self, spider):

        # Close cursor & connection to database
```

```
        self.cur.close()
        self.connection.close()
```

最后,要激活项目管道,需要将其包含在 settings.py 文件中:

```
# settings.py

ITEM_PIPELINES = {
   'postgres_demo.pipelines.PostgresDemoPipeline': 300,
}
```

现在,当我们运行名言爬虫时,PostgresDemoPipeline 将把所有抓来的项存储在数据库中。

如果没有 SQL 数据库查看器,则可以使用 DBeaver(https://dbeaver.io/)。

好的,现在有了一个项目管道,可以将所有抓来的项目保存到 Postgres 数据库中。然而,如果只想保存以前没有抓到的新数据呢?

可以很容易地重新配置管道,让它在再次插入之前检查数据库中的项是否已经在数据库中。为此,将在 pipelines.py 文件中创建一个名为 PostgresNoDuplicatesPipeline 的新管道,并更改 process_item 方法,使其只向数据库中插入新数据。它将首先在数据库中查找 item['text'],只有当它不在时才会插入新项。

```
# pipelines.py

import psycopg2

class PostgresNoDuplicatesPipeline:

    def __init__(self):
        # Connection Details
        hostname = 'localhost'
        username = 'postgres'
        password = '******' # your password
        database = 'quotes'

        # Create/Connect to database
        self.connection = psycopg2.connect(host=hostname, user=username, password=password, dbname=database)

        # Create cursor, used to execute commands
        self.cur = self.connection.cursor()

        # Create quotes table if none exists
        self.cur.execute("""
```

```python
        CREATE TABLE IF NOT EXISTS quotes(
            id serial PRIMARY KEY,
            content text,
            tags text,
            author VARCHAR(255)
        )
        """)

    def process_item(self, item, spider):

        # Check to see if text is already in database
        self.cur.execute("select * from quotes where content = %s", (item['text'],))
        result = self.cur.fetchone()

        # If it is in DB, create log message
        if result:
            spider.logger.warn("Item already in database: %s" % item['text'])


        # If text isn't in the DB, insert data
        else:

            # Define insert statement
            self.cur.execute(""" insert into quotes (content, tags, author) values (%s,%s,%s)""", (
                item["text"],
                str(item["tags"]),
                item["author"]
            ))

            # Execute insert of data into database
            self.connection.commit()
        return item


    def close_spider(self, spider):

        # Close cursor & connection to database
        self.cur.close()
        self.connection.close()
```

要激活此管道，还需要更新 settings.py 以使用 PostgresNoDuplicatesPipeline，而不是以前的 PostgresDemoPipeline 管道：

```
# settings.py

ITEM_PIPELINES = {
    #'postgres_demo.pipelines.PostgresDemoPipeline': 300,
    'postgres_demo.pipelines.PostgresNoDuplicatesPipeline': 300,
}
```

现在，当我们运行名言爬虫时，管道将只存储数据库中尚未存在的新数据。

5.6 Scrapyd：部署、调度和运行 Scrapy 爬虫

Scrapyd 允许我们在服务器上部署 Scrapy 爬虫并使用 JSON API 远程运行它们的应用程序。Scrapyd 允许您：

（1）运行 Scrapy 作业。
（2）暂停和取消 Scrapy 作业。
（3）管理 Scrapy 项目 / 爬虫版本。
（4）远程访问 Scrapy 日志。

使用 Scrapyd，您可以使用现成的 Scrapyd 管理工具从一个中心点管理多台服务器。

设置 Scrapyd 既快速又简单，可以在本地或服务器上运行它。

第一步是安装 Scrapyd：

```
pip install scrapyd
```

然后使用以下命令启动服务器：

```
scrapyd
```

这将启动 Scrapyd。可以在浏览器中打开网址 http://localhost:6800/，显示 Scrapyd 管理界面。

要使用 Scrapyd 运行作业，首先需要把 Scrapy 项目部署到 Scrapyd 服务器。为此，有一个名为 scrapyd-client 的易于使用的库，它使此过程非常简单。

首先安装 scrapyd-client：

```
pip install scrapyd-client
```

安装后，导航到要部署的 Scrapy 项目并打开 scrapyd.cfg 文件，该文件应位于项目的根目录中。然后应该会看到类似这样的内容，其中的 "demo" 文本用 Scrapy 项目名称替换：

```
# scrapy.cfg
```

```
[settings]
default = demo.settings

[deploy]
#url = http://localhost:6800/
project = demo
```

在这里，scrapyd.cfg 配置文件定义了 Scrapy 项目应该部署到的端点。如果我们想将其部署到本地运行的 Scrapyd 服务器，只需取消注释 url 值。

```
# scrapy.cfg

[settings]
default = demo.settings

[deploy]
url = http://localhost:6800/
project = demo
```

然后在 Scrapy 项目的根目录中运行以下命令：

```
scrapyd-deploy default
```

这将使 Scrapy 项目部署到本地运行的 Scrapyd 服务器。

上面的示例是最简单的实现，并假设您只是将 Scrapy 项目部署到本地 Scrapyd 服务器。但是，也可以自定义或添加多个部署端点到 scrapyd.cfg 文件。

例如，可以定义本地和生产端点：

```
# scrapy.cfg

[settings]
default = demo.settings

[deploy:local]
url = http://localhost:6800/
project = demo

[deploy:production]
url = <MY_IP_ADDRESS>
project = demo
```

并使用以下命令在本地或生产环境中部署 Scrapy 项目：

```
# Deploy locally
scrapyd-deploy local
```

```
# Deploy to production
scrapyd-deploy production
```

Scrapyd 带有一个最小的 Web 界面，可以通过 http://localhost:6800/ 访问。但是，这个界面只是对 Scrapyd 服务器上运行的内容的基本概述，并且不允许您控制部署到 Scrapyd 服务器的爬虫。

要在您的 Scrapyd 服务器上安排、运行、取消作业，需要使用它提供的 JSON API。根据端点，API 支持 GET 或 POST HTTP 请求，例如：

```
>curl http://localhost:6800/daemonstatus.json
{ "status": "ok", "running": "0", "pending": "0", "finished": "0", "node_name":
"DESKTOP-67BR2" }
```

API 具有表 5-1 所示端点。

表 5-1 端点表

端点	描述
daemonstatus.json	检查 Scrapyd 服务器的状态
addversion.json	向项目添加版本，如果项目不存在则创建项目
schedule.json	安排要运行的作业
cancel.json	取消作业。如果作业处于待处理状态，它将被删除；如果作业正在运行，则作业将被关闭
listprojects.json	返回上传到 Scrapyd 服务器的项目列表
listversions.json	返回可用于请求项目的版本列表
listspiders.json	返回可用于请求项目的爬虫列表
listjobs.json	返回所请求项目的待处理、运行和已完成的作业的列表
delversion.json	删除项目版本。如果项目只有一个版本，则也删除该项目
delproject.json	删除项目和所有关联的版本

我们可以使用 Python Requests 或任何其他 HTTP 请求库与这些端点进行交互，或者可以使用 python-scrapyd-api，一个用于 Scrapyd API 的 Python 包装器。

python-scrapyd-api 围绕 Scrapyd JSON API 提供了一个干净且易于使用的 Python 包装器，它可以简化代码。

首先，需要安装它：

```
pip install python-scrapyd-api
```

然后在代码中，我们需要导入库并将其配置为可以与我们的 Scrapyd 服务器交互，方法是向其传递 Scrapyd IP 地址。

```
>>> from scrapyd_api import ScrapydAPI
>>> scrapyd = ScrapydAPI('http://localhost:6800')
```

从这里，我们可以使用内置方法与 Scrapyd 服务器交互。

返回上传到 Scrapyd 服务器的项目列表。

```
>>> scrapyd.list_projects()
[u'demo', u'quotes_project']
```

输入项目名称，它将返回可用于所请求项目的爬虫列表。

```
>>> scrapyd.list_spiders('project_name')
[u'raw_spider', u'js_enhanced_spider', u'selenium_spider']
```

通过指定项目和爬虫名称来运行 Scrapy 爬虫。

```
>>> scrapyd.schedule('project_name', 'spider_name')
# Returns the Scrapyd job id.
u'14a6599ef67111e38a0e080027880ca6'
```

通过发送项目名称和 job_id 取消正在运行的作业。

```
>>> scrapyd.cancel('project_name', '14a6599ef67111e38a0e080027880ca6')
# Returns the "previous state" of the job before it was cancelled: 'running' or 'pending'.
'running'
```

发送后，它将返回作业被取消之前的状态。可以通过检查作业状态来验证作业是否实际被取消。

```
>>> scrapyd.job_status('project_name', '14a6599ef67111e38a0e080027880ca6')
# Returns 'running', 'pending', 'finished' or '' for unknown state.
'finished'
```

5.7 Scrapy Cloud 托管爬虫

Scrapy Cloud 是一个可扩展的云托管解决方案，用于运行和调度 Scrapy 爬虫，由 Zyte 创建。

在本节中，将部署 Scrapy 爬虫 booksbot（https://github.com/scrapy/booksbot）。要使用它，只需将其克隆到您的机器上即可。

```
git clone https://github.com/scrapy/booksbot.git
cd booksbot
```

开始使用 Scrapy Cloud 非常简单。

首先在这里创建一个免费账户 Scrapy Cloud，然后登录后单击"启动一个新项目"，给项目起个名字。

一旦项目创建完成，有两种方法可以将 Scrapy 爬虫部署到 Scrapy Cloud：

（1）通过命令行；

（2）通过 GitHub 集成。

使用 shub 命令行工具，可以从命令行将爬虫直接部署到 Scrapy Cloud。

首先在系统上安装 shub：

```
pip install shub
```

然后，通过在命令行中运行 shub 登录，将 shub 客户端连接到 Scrapy Cloud 项目，并在提示时输入 Scrapy Cloud API 键。

```
shub login
API key: YOUR_API_KEY
```

可以在"代码和部署"页面上找到 API 密钥，然后，要将 Scrapy 项目部署到 Scrapy Cloud，请运行 shub-deploy 命令，后跟项目 ID：

```
shub deploy PROJECT_ID
```

可以在"代码和部署"页面或项目 URL 中找到项目的 ID。

```
https://app.zyte.com/p/PROJECT_ID/jobs
```

如果成功，将在爬虫选项卡中看到可用的爬虫。

然后，就可以直接从命令行在 Scrapy Cloud 上运行抓取作业：

```
$ shub schedule books
Spider books scheduled, watch it running here:
https://app.zyte.com/p/26731/job/1/8
```

另一种选择是将 Scrapy Cloud 直接连接到 GitHub 账户，并直接从 GitHub 部署爬虫。

在"代码和部署"页面上，选择"连接到 GitHub"选项，然后按照说明进行操作。如果以前没有将 Zyte 连接到 GitHub 账户，那么可能会被要求授权 Zyte 访问存储库。接下来，系统将提示选择希望 Scrapy Cloud 连接到哪个存储库。

默认情况下，当将 Scrapy Cloud 连接到 GitHub 存储库时，它被配置为自动部署推送到 GitHub 存储库的任何更改。但是，如果你愿意，可以将其切换到手动部署模式，并手动部署对爬虫的更改。如果将其保留在"自动部署"模式，则要开始第一次部署时，请单击"部署分支"按钮。

如果成功，您将在爬虫选项卡中看到可用的爬虫。

一旦部署了爬虫，就可以在 Scrapy Cloud 上运行抓取作业，这非常简单。只需转到爬虫控制面板，选择要运行的爬虫，然后单击运行。可以选择在作业运行之前向作业添加任何参数、标记或额外的 Scrapy 单元。

一旦满意了，然后单击运行，Scrapy Cloud 就会将要运行的作业排队。

现在，当转到作业面板时，将看到作业是否已排队、正在运行或已完成。以及一些概述统计信息，如运行时、抓取的项目、错误等。

Scrapy Cloud 最有用的功能是定时任务功能，它允许您安排爬虫在未来定期运行。Scrapy Cloud 使用类似于 crontab 的调度器，因此可以将爬虫安排为每分钟、每小时、每天、每周或每月运行一次。

要使用计划功能，请转到"定时任务"面板，然后单击"添加定时任务"。在这里，系统会提示选择要安排的爬虫，它应该何时运行，以及任何额外的设置，如优先级、标记和参数。保存后，此爬虫将自动按选择的时间间隔运行。

5.8 Twisted 框架

Scrapy 是用 Twisted 编写的，这是一个流行的 Python 事件驱动网络框架，因此，Scrapy 是使用非阻塞（也称为异步）代码来实现并发的。

首先，熟悉 Twisted 背后的计算模型（异步编程）很重要。让我们从回顾其他两种传统的计算模型开始，以便与它们进行比较：单线程模型和多线程模型。

传统的单线程（同步）模型非常简单：一次只执行一个任务，直到前一个任务完成后才能开始新任务。

在多线程模型中，每个任务都由一个单独的操作系统线程执行，操作系统可以随时用另一个线程替换正在运行的任务。在具有多个内核的系统上，不同的线程可以真正并发运行，也可以在单个内核上交错运行。但需要注意的是，在 Python 中，由于全局解释器锁（GIL），多线程应用程序永远不会真正并发运行。

在异步模型中，任务在单个线程中相互交错。遵循此模型的应用程序的主要特点是，当正在运行的任务将被阻塞时，应用程序将继续执行另一个任务，从而最大限度地减少整个应用程序被阻塞的时间。在这个模型中，运行中的任务在完成或将被阻塞时被另一个任务替换。

如果一个应用程序由一组独立的任务组成，并且有相当多的阻塞，那么使用该模型会使应用程序受益。任务阻塞的最常见原因是等待 I/O 完成，例如从网络或文件系统中读取或写入。

Twisted 有两个主要组件：Reactor 和 Deferred。

Reactor 对事件做出反应并安排任务，这就是为什么它被称为反应器或事件循环。Reactor 的工作是管理等待执行的任务池。当正在运行的任务通过启动异步操作将控制权交给 Reactor 时，Reactor 负责将任务置于等待状态，设置一种机制，在结果可用时调用与任务关联的结果处理程序，并继续执行另一个任务。

Twisted 的另一个重要组成部分是 Deferred。Deferred 封装异步任务，类似于其他框架中的 Promise 或 Future。Twisted 中的异步函数返回一个 Deferred，Deferred 用于控制任务的执行，并在结果可用时访问该结果。

为了与 Deferred 交互，当异步任务的结果可用时，我们可以附加一系列要调用的函数。这一系列函数称为回调或回调链。如果异步任务中存在错误，我们也可以附加一个要调用的函数列表，称为 errback 链。当结果可用时调用第一个回调，如果出现错误，则调用第一个 errback。

让我们从一个简单的示例开始，获取 URL 中的内容并打印返回的 HTTP 代码。为此，我们将使用 Treq，这是一个基于 Twisted 之上编写的库。

```python
from twisted.internet.task import react
import treq

def get_url(url):

    def handle_response(resp):
        print("Got response code %d from %s" % (resp.code, url))

    def handle_failure(failure):
        print("Something failed: %s" % failure.getErrorMessage())

    print("Getting URL %s" % url)
    d = treq.get(url)
    d.addCallback(handle_response)
    d.addErrback(handle_failure)
    return d

def main(reactor, *args):
    url = 'http://baidu.com'
    d = get_url(url)
    return d

react(main)
```

让我们按部分分析此代码：

（1）react 是 Twisted 提供的一个实用函数，它启动 reactor，执行提供的函数（在本例中为 main），并在所有任务完成后停止 reactor。

（2）main 函数负责使用特定的 url 调用 get_url。

（3）get_url 调用 treq.get，获取请求的 URL 上的内容，因为它是一个异步函数，treq.get 返回一个 Deferred。

（4）我们在返回的 Deferred 中添加了一个回调函数，用于打印响应代码，并在出现错误时添加一个 errback 函数，用于打印错误。

第一个示例与普通的单线程程序非常相似，让我们稍微修改第一个示例以同时获得多个 URL，从而了解更多的 Twisted。我们只需要为此修改主函数：

```
from twisted.internet.task import defer

def main(reactor, *args):
    urls = ['http://github.com', 'http://www.example.com', 'http://www.sohu.com']
    ds = map(get_url, urls)
    d = defer.gatherResults(ds)
    return d
```

这里我们为多个 url 调用 get_url() 函数，每次调用都会得到一个 Deferred。然后，我们使用 Twisted 函数 gatherResults() 来创建一个新的 Deferred，当所有提供的 Deferreds 都被激发时，它就会被激发。

5.9 本章小结

本章介绍了 Scrapy 抓取静态网站和动态网站，以及将抓取的数据保存到数据库。

Scrapy 是一个用 Python 编写的免费开源网络爬虫框架。最初设计用于网络抓取，它也可用于使用 API 提取数据或用作通用网络爬虫。它目前由 Zyte 公司维护，前身为 Scrapinghub 公司，这是一家网络抓取开发和服务公司。

第 6 章 分布式爬虫开发

本章介绍如何实现分布式爬虫,包括 Web 抓取、提取内容,以及用容错方式以可伸缩性存储内容。

Celery 是基于分布式消息传递的异步任务队列 / 作业队列。我们将使用它实现分布式爬虫。

6.1 简单的 Celery 任务

Celery 的 Redis 支持需要其他依赖项。使用 Celery [redis] 捆绑包一次性安装 Celery 和依赖项的步骤如下。

配置 Redis 数据库的位置:

```
BROKER_URL = 'redis://localhost:6379/0'
```

URL 的格式应为:

```
redis://:password@hostname:port/db_number
```

创建文件 tasks.py:

```
from celery import Celery

BROKER_URL = 'redis://localhost:6379/0'
app = Celery('tasks', broker=BROKER_URL)

@app.task
def add(x, y):
    return x + y
```

Celery 的第一个参数是当前模块的名称,这样可以自动生成名称。第二个参数是 broker 关键字,它指定消息代理的 URL。

启动 Redis 服务器后,运行 Celery 工作者:

```
celery -A tasks worker --loglevel=info  -P eventlet -c 10
```

要调用任务，可使用 delay() 方法：

```
>>> from tasks import add
>>> add.delay(4, 4)
```

调用任务会返回一个 AsyncResult 实例，该实例可以检查任务的状态、等待任务完成或获取其返回值（如果任务失败，它将获取异常和回溯）。

为了跟踪任务的状态，Celery 需要将状态存储或发送到某处。使用 Redis 作为结果后端的代码如下：

```
from celery import Celery

BROKER_URL = 'redis://localhost:6379/0'
BACKEND_URL = 'redis://localhost:6379/1'
app = Celery('tasks', broker=BROKER_URL, backend=BACKEND_URL)

@app.task
def add(x, y):
    return x + y
```

现在配置好结果后端后，再次调用该任务。这一次保留从任务返回的 AsyncResult 实例：

```
>>> result = add.delay(4, 4)
```

ready() 方法返回任务是否已完成处理：

```
>>> result.ready()
False
```

可以等待结果完成，但很少使用，因为它会将异步调用转换为同步调用：

```
>>> result.get(timeout=1)
8
```

6.2 从任务进行分布式抓取

下一步是将 Celery 任务与爬虫进程连接起来。

```
from celery import Celery
import requests
from bs4 import BeautifulSoup
from urllib.parse import urljoin

app = Celery('tasks', broker_url='redis://127.0.0.1:6379/1')
```

```python
@app.task
def crawl(url):
    html = get_html(url)
    soup = BeautifulSoup(html, 'html.parser')
    links = extract_links(url, soup)
    print(links)

def get_html(url):
    try:
            response = requests.get(url)
            return response.content
    except Exception as e:
            print(e)

    return ''

def extract_links(url, soup):
    return list({
            urljoin(url, a.get('href'))
            for a in soup.find_all('a')
            if a.get('href') and not(a.get('rel') and 'nofollow' in a.get('rel'))
    })

starting_url = 'https://scrapeme.live/shop/page/1/'
crawl.delay(starting_url)
```

我们需要保存所有这些数据：访问过的页面、当前正在爬取的页面、保留"待访问"所列表以及稍后存储的一些内容。

这里将使用 Redis 来避免重新爬取，而不是直接排队到 Celery，并且只将 URL 排队一次。这里不会深入讨论 Redis 的更多细节，但会使用列表、集合和散列。

先创建一个名为 crawling:to_visit 的列表，并推送起始 URL。然后，我们将进入一个循环，该循环将查询该列表中的项目并阻塞其一分钟，直到项目就绪。当检索到项目时，我们调用爬取函数，将其执行排队。

```python
from redis import Redis
from tasks import crawl

connection = Redis(db=1)
starting_url = 'https://scrapeme.live/shop/page/1/'

connection.rpush('crawling:to_visit', starting_url)
```

```
while True:
    # timeout after 1 minute
    item = connection.blpop('crawling:to_visit', 60)
    if item is None:
            print('Timeout! No more items to process')
            break

    url = item[1].decode('utf-8')
    print('Pop URL', url)
    crawl.delay(url)
```

我们将使用集合跟踪所有排队和访问的集合，并在它们的总和超过允许的最大值时退出。

```
from redis import Redis
# ...
connection = Redis(db=1)

@app.task
def crawl(url):
    connection.sadd('crawling:queued', url) # add URL to set
    html = get_html(url)
    soup = BeautifulSoup(html, 'html.parser')
    links = extract_links(url, soup)
    for link in links:
            if allow_url_filter(link) and not seen(link):
                    print('Add URL to visit queue', link)
                    add_to_visit(link)

    # atomically move a URL from queued to visited
    connection.smove('crawling:queued', 'crawling:visited', url)

def allow_url_filter(url):
    return '/shop/page/' in url and '#' not in url

def seen(url):
    return connection.sismember('crawling:visited', url) or connection.sismember('crawling:queued', url)

def add_to_visit(url):
    # LPOS command is not available in Redis library
    if connection.execute_command('LPOS', 'crawling:to_visit', url) is None:
            connection.rpush('crawling:to_visit', url) # add URL to the end of the list
```

执行后，所有内容都将在 Redis 中，因此再次运行将无法按预期工作，我们需要手动清理爬虫队列，我们可以通过使用 redis-cli 或类似于 GUI 的 redis 命令来实现。有一些命令用于删除键（即 DEL crawling:to_visit）或刷新数据库。

在项目发展之前，我们将开始分离概念。我们已经有了两个文件：tasks.py 和 main.py。我们将创建另外两个文件来托管爬虫相关函数（crawler.py）和数据库访问（repo.py）。repo.py 的部分代码如下。

```python
from redis import Redis

connection = Redis(db=1)

to_visit_key = 'crawling:to_visit'
visited_key = 'crawling:visited'
queued_key = 'crawling:queued'

def pop_to_visit_blocking(timeout=0):
    return connection.blpop(to_visit_key, timeout)

def count_visited():
    return connection.scard(visited_key)

def is_queued(value):
    return connection.sismember(queued_key, value)
```

crawler 文件具有抓取、提取链接等功能。

我们需要一些方法来提取和存储内容，并只向队列添加特定的链接子集。我们需要一个新的概念：默认解析器（parsers/defaults.py）。

```python
import repo

def extract_content(url, soup):
    return soup.title.string # extract page's title

def store_content(url, content):
    # store in a hash with the URL as the key and the title as the content
    repo.set_content(url, content)

def allow_url_filter(url):
    return True # allow all by default

def get_html(url):
    # ... same as before
```

在 repo.py 文件中：

```
content_key = 'crawling:content'
# ..
def set_content(key, value):
    connection.hset(content_key, key=key, value=value)
```

解析器允许我们抽象链接和内容提取,它是一组作为参数传递的函数,而不是在爬虫中硬编码。现在我们可以用 import 替换对这些函数的调用。

为了完全抽象,我们需要一个生成器或工厂。我们将创建一个新文件来承载它——parserlist.py。为了简化一点,我们允许每个域有一个自定义解析器。该演示包括两个测试域:scrapeme.live 和 quotes.tocrap.com。

目前还没有对每个域进行任何操作,因此我们将使用默认的解析器。

```
from urllib.parse import urlparse
from parsers import defaults

parsers = {
    'scrapeme.live': defaults,
    'quotes.toscrape.com': defaults,
}

def get_parser(url):
    hostname = urlparse(url).hostname # extract domain from URL
    if hostname in parsers:
            # use the dict above to return the custom parser if present
            return parsers[hostname]
    return defaults
```

现在,我们可以使用新的解析器修改任务。

```
@app.task
def crawl(url):
    parser = get_parser(url) # get the parser, either custom or the default one
    html = parser.get_html(url)
    # ...
    for link in links:
            if parser.allow_url_filter(link) and not seen(link):
                    # ...
```

自定义解析器将首先使用 scrapme 作为示例。

本部分要求了解页面及其 HTML。总之,我们将获得产品列表中每个项目的产品 id、名称和价格,然后使用 id 作为键将其存储在一个集合中。至于允许的链接,只有用于分页的链接将通过过滤。

```
import json
```

```python
import defaults
import repo

def extract_content(url, soup):
    return [{
        'id': product.find('a',
                attrs={'data-product_id': True})['data-product_id'],
        'name': product.find('h2').text,
        'price': product.find(class_='amount').text
    } for product in soup.select('.product')]

def store_content(url, content):
    for item in content:
        if item['id']:
            repo.set_content(item['id'], json.dumps(item))

def allow_url_filter(url):
    return '/shop/page/' in url and '#' not in url

def get_html(url):
    return defaults.get_html(url)
```

在名言网站中,我们需要以不同的方式处理它,因为每个名言没有 id。我们将为列表中的每个条目提取作者和名言,然后,在 store_content 函数中,我们将为每个作者创建一个列表并添加该名言。Redis 在必要时处理列表的创建。

```python
def extract_content(url, soup):
    return [{
        'quote': product.find(class_='text').text,
        'author': product.find(class_='author').text
    } for product in soup.select('.quote')]

def store_content(url, content):
    for item in content:
        if item['quote'] and item['author']:
            list_key = f"crawling:quote:{item['author']}"
            repo.add_to_list(list_key, item['quote'])
```

通过最后几次更改,我们引入了易于扩展的自定义解析器。添加新站点时,我们必须为每个新域名创建一个文件,并在 parserlist.py 中增加一行引用该文件。

对于本地测试,我们可以启动两个不同的工作者 celery -A tasks worker --concurrency=20 -n worker1 和 ... -n worker2,但这不是一个实际的分布式网络爬虫设计。

重要的是要注意,工作者的名字很重要,特别是在同一台机器上启动多个工作者时。

如果在不更改工作者名字的情况下执行上述命令两次，Celery将无法正确地识别他们，因此，将第二个作为 -n worker2 启动。

如果项目增长，唯一的节点将是瓶颈。要正确制作分布式爬虫，我们需要多个节点，它们中的每一个都会运行相同的代码，并且可以访问代理——在我们的例子中，就是Redis。Celery处理工作者并分配负载。

6.3 本章小结

本章介绍了使用异步任务队列系统Celery实现分布式爬虫的方法。

第 7 章　开发网络爬虫用户界面

本章首先介绍开发网络爬虫图形用户界面所使用的 Tkinter 模块，然后介绍应用 Tkinter 实现网络爬虫 GUI。

7.1　Tkinter 简介

Tkinter 是用于创建 GUI 应用程序的内置 Python 模块，它是在 Python 中创建 GUI 应用程序最常用的模块之一，因为它简单易用。我们不必担心 Tkinter 模块的单独安装，因为它已经随 Python 一起提供了。它为 Tk GUI 工具包提供了一个面向对象的界面。

Tkinter 是创建各种图形用户界面的有用工具，包括窗口、对话框和自定义小部件，它特别适合于构建桌面应用程序和向命令行程序添加 GUI。

基本 Tkinter 小部件如表 7-1 所示。

表 7-1　基本 Tkinter 小部件

小 部 件	描　　述
Label	用于在屏幕上显示文本或图像
Button	用于向应用程序添加按钮
Canvas	用于绘制图片和其他布局，如文本、图形等
ComboBox	包含从可用选项列表中选择的向下箭头
CheckButton	向用户显示许多选项作为切换按钮，用户可以从中选择任意数量的选项
RadioButton	用于实现多种选择中的一种，因为它只允许选择一个选项
Entry	用于输入用户的单行文本输入
Frame	用作容器来保存和组织小部件
Message	工作原理与标签相同，并涉及多行和不可编辑的文本
Scale	用于提供一个图形滑块，允许从该比例中选择任何值
Scrollbar	用于向下滚动的内容，它提供了一个滑动控制器

续表

小部件	描述
SpinBox	允许用户从给定的一组值中进行选择
Text	允许用户编辑多行文本并设置其显示方式
Menu	用于创建应用程序使用的各种菜单

使用 Tkinter 的示例如下：

```
from tkinter import *
from tkinter.ttk import *

# writing code needs to
# create the main window of
# the application creating
# main window object named root
root = Tk()

# giving title to the main window
root.title("First_Program")

# Label is what output will be
# show on the window
label = Label(root, text="Hello World !").pack()

# calling mainloop method which is used
# when your application is ready to run
# and it tells the code to keep displaying
root.mainloop()
```

使用 Tkinter 创建按钮：

```
# import everything from tkinter module
from tkinter import *

# create a tkinter window
root = Tk()

# Open window having dimension 100x100
root.geometry('100x100')

# Create a Button
btn = Button(root, text = 'Click me !', bd = '5',
                        command = root.destroy)
```

```
# Set the position of button on the top of window.
btn.pack(side = 'top')

root.mainloop()
```

使用 tk 主题控件（tkinter.ttk）创建按钮，这将为您提供现代图形的效果。

```
# import tkinter module
from tkinter import *

# Following will import tkinter.ttk module and
# automatically override all the widgets
# which are present in tkinter module.
from tkinter.ttk import *

# Create Object
root = Tk()

# Initialize tkinter window with dimensions 100x100
root.geometry('100x100')

btn = Button(root, text = 'Click me !',
             command = root.destroy)

# Set the position of button on the top of window
btn.pack(side = 'top')

root.mainloop()
```

注意：请参阅两个代码的输出，因为 tkinter.ttk 不支持边界，所以 tkinter.ttk 代码输出中不存在 BORDER。此外，当您将鼠标悬停在两个按钮上时，ttk.Button 将改变其颜色并变为浅蓝色（效果可能会因操作系统而异），因为它支持现代图形，而对于简单的按钮，它不会改变颜色，因为它不支持现代图形。

Button 小部件是 Tkinter 库中最基本的小部件之一，这是用户与 Tkinter 窗口交互的一种方式。单击按钮后，程序将触发一个操作。

Tkinter 允许使用各种不同的函数和参数自定义这些按钮。此类自定义的例子包括颜色、字体类型、字体大小、图标和图像。

Button 小部件语法：

```
object = Button(master, options.....)
```

按钮选项如下。

- **activebackground**：光标位于按钮上方时的背景颜色。

- activefrontground：光标位于按钮上方时的前景颜色。
- bg：按钮的背景色。
- bd：以像素为单位的边框大小。默认值为2。
- command：单击按钮时要执行的命令。通常设置为函数调用。
- fg：前景颜色。
- font：按钮的文本字体。
- height：按钮的高度。
- highlightcolor：小部件聚焦时的文本颜色。
- image：按钮上显示的图像。默认情况下，图像将替换文本。
- justify：更改文本的对齐方式。可以设置为 LEFT、CENTER 或 RIGHT。
- padx：在文本的左侧和右侧填充。
- pady：在文本上方和下方填充。
- relief：指定边框的类型。此选项的默认值为 FLAT，其他选项包括 SUNKEN、RAISED、GROVE 和 RIDGE。
- state：默认值为 NORMAL。DISABLED（禁用）会使按钮变灰并处于非活动状态。ACTIVE 是鼠标悬停在按钮上时的状态。
- underline：默认值为 -1。为了给按钮文本添加下画线，可以设置此选项。
- width：按钮的宽度。
- wrappelength：如果该值设置为正数，则文本行将被换行以适应此长度。

下面的代码将创建一个带有按钮的 GUI，该按钮将向控制台显示文本"Hello World"。

```
from tkinter import *

def set():
    print("Hello World")

root = Tk()
root.geometry("200x150")

frame = Frame(root)
frame.pack()
button = Button(frame, text = "Button1", command = set)
button.pack()

root.mainloop()
```

Tkinter 标签是一个小部件，用于实现可以放置文本或图像的显示框。开发人员可以随时更改此小部件显示的文本。它还用于执行在文本部分添加下画线，跨多行显示文本这样的任务。需要注意的是，标签一次只能使用一种字体来显示文本。要使用标签，只需指定要在其中显示的内容（可以是文本、位图或图像）。

语法：

```
w = Label ( master, option, … )
```

参数如下。

- master：表示父窗口。
- options：下面是这个小部件最常用的选项列表。这些选项可用作以逗号分隔的键值对。

各种选项描述如下。

- anchor：如果小部件的空间大于文本所需的空间，则此选项用于控制文本的位置。默认值为 anchor=CENTER，将文本置于可用空间的中心。
- bg：此选项用于设置标签和指示器后面显示的正常背景色。
- height：此选项用于设置新框架的垂直尺寸。
- width：标签的宽度，以字符为单位（不是像素！）。如果未设置此选项，则标签的大小将适合其内容。
- bd：此选项用于设置指示器周围边框的大小。默认的 bd 值为 2 像素。
- font：如果您在标签中显示文本，则字体选项用于指定标签中的文本将以何种字体显示。
- cursor：用于指定当鼠标移动到标签上时要显示的光标。默认值是使用标准光标。
- textvariable：顾名思义，它与标签的 Tkinter 变量（通常是 StringVar）关联。如果变量已更改，则标签文本将更新。
- bitmap：用于将位图设置为指定的图形对象，以便标签可以表示图形而不是文本。
- fg：如果在此标签中显示文本或位图，则此选项指定文本的颜色。如果要显示位图，则这是位图中 1 位位置处出现的颜色。
- image：此选项用于在标签小部件中显示静态图像。
- padx：在小部件中为文本的左侧和右侧添加额外的空间。默认值为 1。
- pady：在小部件的文本上方和下方添加额外的空间。默认值为 1。
- justify：此选项用于定义如何对齐多行文本。使用 LEFT、RIGHT 或 CENTER 作为其值。请注意，要在小部件中定位文本，请使用锚定选项。对齐的默认值为 CENTER。

- relief：此选项用于指定标签周围装饰边框的外观。此选项的默认值为 FLAT。
- underline：通过将此选项设置为 n，可以在文本的第 n 个字母下方显示下画线 (_)，从 0 开始计数。默认值为 underline=-1，这意味着没有下画线。
- wrappelength：不是只有一行作为标签文本，而是可以拆分为数行，其中每行都有为此选项指定的字符数。

示例代码如下：

```python
from tkinter import *

top = Tk()
top.geometry("450x300")

# the label for user_name
user_name = Label(top,
                  text = "Username").place(x = 40,
                                           y = 60)


# the label for user_password
user_password = Label(top,
                      text = "Password").place(x = 40,
                                               y = 100)


submit_button = Button(top,
                       text = "Submit").place(x = 40,
                                              y = 130)


user_name_input_area = Entry(top,
                             width = 30).place(x = 110,
                                               y = 60)


user_password_entry_area = Entry(top,
                                 width = 30).place(x = 110,
                                                   y = 100)


top.mainloop()
```

Canvas 小部件允许我们在应用程序上显示各种图形，它可以用来绘制简单的形状到复杂的图形。我们还可以根据需要显示各种自定义小部件。

语法：

```
C = Canvas(root, height, width, bd, bg, ..)
```

可选参数如下。

- root：根窗口。
- height：画布小部件的高度。
- width：画布小部件的宽度。
- bg：画布的背景色。
- bd：画布窗口的边框。
- scrollregion：一个元组（w，n，e，s），定义画布可以滚动的区域大小，其中w是左侧，n是顶部，e是右侧，s是底部。
- highlightcolor：焦点高亮显示的颜色。
- cursor：可以定义为画布的光标，可以是圆、箭头等。
- confine：决定是否可以在滚动区域之外访问画布。
- relief：边框的类型，可以是SUNKEN、RAISED、GROVE和RIDGE。

一些常见的绘图方法：

- 创建椭圆

```
oval = C.create_oval(x0, y0, x1, y1, options)
```

- 创建直线

```
line = C.create_line(x0, y0, x1, y1, ..., xn, yn, options)
```

- 创建多边形

```
oval = C.create_polygon(x0, y0, x1, y1, ...xn, yn, options)
```

- 创建图像

```
img = canvas.create_image(x, y, image=img_path)
```

简单形状的绘图的代码如下：

```
from tkinter import *

root = Tk()

C = Canvas(root, bg="yellow",
          height=250, width=300)

line = C.create_line(108, 120,
                    320, 40,
                    fill="green")

oval = C.create_oval(80, 30, 140,
                    150,
```

```
                    fill="blue")
C.pack()
mainloop()
```

Entry 小部件是一个用于输入或显示单行文本的小部件。

语法：

```
entry = tk.Entry(parent, options)
```

参数描述如下。

（1）parent：要显示的小部件的父级窗口或框架。

（2）optiens：Entry 小部件提供的各种选项。各种选项描述如下。

- bg：标签和指示器后面显示的背景色。
- bd：指示器周围边框的大小。默认值为 2 像素。
- font：用于文本的字体。
- fg：用于渲染文本的颜色。
- justify：如果文本包含多行，则此选项控制文本的对齐方式。可选值为 CENTER、LEFT 或 RIGHT。
- relief：边框的类型。relief 默认值是 FLAT。您可以将此选项设置为任何其他样式，如 SUNKEN、RIGID、RAISED、GROVE。
- show：通常，用户输入的字符会出现在 entry 中。如果要创建一个 .password.entry，且将每个字符作为星号进行响应，请设置 show="*"。
- textvariable：为了能够从 Entry 小部件中检索当前文本，必须将此选项设置为 StringVar 类的实例。

Entry 小部件提供的各种方法如下。

- get()：以字符串形式返回条目的当前文本。
- delete()：从小部件中删除字符。
- insert（index，'name'）：在给定索引的字符之前插入字符串"name"。

示例代码如下：

```
# Program to make a simple
# login screen

import tkinter as tk

root=tk.Tk()
```

```python
    # setting the windows size
    root.geometry("600x400")

    # declaring string variable
    # for storing name and password
    name_var=tk.StringVar()
    passw_var=tk.StringVar()


    # defining a function that will
    # get the name and password and
    # print them on the screen
    def submit():

        name=name_var.get()
        password=passw_var.get()

        print("The name is : " + name)
        print("The password is : " + password)

        name_var.set("")
        passw_var.set("")


    # creating a label for
    # name using widget Label
    name_label = tk.Label(root, text = 'Username', font=('calibre',10, 'bold'))

    # creating a entry for input
    # name using widget Entry
    name_entry = tk.Entry(root,textvariable = name_var, font=('calibre',10,'normal'))

    # creating a label for password
    passw_label = tk.Label(root, text = 'Password', font = ('calibre',10,'bold'))

    # creating a entry for password
    passw_entry=tk.Entry(root, textvariable = passw_var, font = ('calibre',10,'normal'), show = '*')

    # creating a button using the widget
    # Button that will call the submit function
    sub_btn=tk.Button(root,text = 'Submit', command = submit)

    # placing the label and entry in
    # the required position using grid
```

```
# method
name_label.grid(row=0,column=0)
name_entry.grid(row=0,column=1)
passw_label.grid(row=1,column=0)
passw_entry.grid(row=1,column=1)
sub_btn.grid(row=2,column=1)

# performing an infinite loop
# for the window to display
root.mainloop()
```

Text 小部件用于用户希望插入多行文本字段的地方。此小部件可用于需要多行文本的各种应用程序，例如消息传递、发送信息或显示信息以及许多其他任务。我们还可以在 Text 小部件中插入图像和链接等媒体文件。

语法：

```
T = Text(root, bg, fg, bd, height, width, font, ..)
```

可选参数描述如下。

- root：根窗口。
- bg：背景色。
- fg：前景色。
- bd：小部件的边框。
- height：小部件的高度。
- width：小部件的宽度。
- font：文本的字体类型。
- cursor：要使用的光标类型。
- insetoffime：光标闪烁关闭的时间（毫秒）。
- insertontime：光标闪烁的时间（毫秒）。
- padx：水平填充。
- pady：垂直填充。
- state：定义小部件是否响应鼠标或键盘的移动。
- highlightthickness：定义焦点高光的厚度。
- insertionwidth：定义插入字符的宽度。
- relief：边框的类型，可以是 SUNKEN、RIGID、RAISED 和 GROVE。
- yscrollcommand：使小部件可垂直滚动。
- xscrollcommand：使小部件可水平滚动。

一些常用方法如下。

- index(index)：获取指定的索引。
- insert(index)：在指定索引处插入字符串。
- see(index)：检查字符串在给定索引处是否可见。
- get(startindex, endindex)：获取给定范围内的字符。
- delete(startindex, endindex)：删除指定范围内的字符。

示例代码如下：

```python
import tkinter as tk
from tkinter import *

root = Tk()

# specify size of window.
root.geometry("250x170")

# create text widget and specify size.
T = Text(root, height=5, width=52)

# create label
l = Label(root, text="Fact of the Day")
l.config(font=("Courier", 14))

Fact = """A man can be arrested in
Italy for wearing a skirt in public."""

# create button for next text.
b1 = Button(root, text="Next", )

# create an Exit button.
b2 = Button(root, text="Exit",
            command=root.destroy)

l.pack()
T.pack()
b1.pack()
b2.pack()

# insert The Fact.
T.insert(tk.END, Fact)

tk.mainloop()
```

Frame 是屏幕上的矩形区域。Frame 可以用作基础类来实现复杂的小部件，它用于组织一组小部件。

使用 Frame 小部件的语法如下所示。

```
w = Frame( master, options)
```

参数描述如下。

- master：此参数用于表示父窗口。
- options：有许多可用的选项，它们可以用作用逗号分隔的键值对。

以下是可用于此小部件常用的选项。

- bg：此选项用于表示标签和指示器后面显示的正常背景色。
- bd：此选项用于表示指示器周围边框的大小，默认值为 2 像素。
- cursor：通过使用此选项，鼠标光标在 Frame 上时将更改为该模式。
- height：新 Frame 的垂直尺寸。
- highlightcolor：此选项用于表示 Frame 具有焦点时焦点高亮显示的颜色。
- highlightthickness：焦点高亮显示的厚度。
- highlightbackground：此选项用于表示 Frame 没有焦点时焦点高亮显示的颜色。
- relief：Frame 边框的类型。其默认值设置为 FLAT。
- width：此选项用于表示 Frame 的宽度。

示例代码如下：

```
from tkinter import *

root = Tk()
root.geometry("300x150")

w = Label(root, text='GeeksForGeeks', font="50")
w.pack()

frame = Frame(root)
frame.pack()

bottomframe = Frame(root)
bottomframe.pack(side=BOTTOM)

b1_button = Button(frame, text="Geeks1", fg="red")
b1_button.pack(side=LEFT)

b2_button = Button(frame, text="Geeks2", fg="brown")
b2_button.pack(side=LEFT)
```

```
b3_button = Button(frame, text="Geeks3", fg="blue")
b3_button.pack(side=LEFT)

b4_button = Button(bottomframe, text="Geeks4", fg="green")
b4_button.pack(side=BOTTOM)

b5_button = Button(bottomframe, text="Geeks5", fg="green")
b5_button.pack(side=BOTTOM)

b6_button = Button(bottomframe, text="Geeks6", fg="green")
b6_button.pack(side=BOTTOM)

root.mainloop()
```

Toplevel 小部件用于在所有其他窗口之上创建窗口。Toplevel 小部件用于向用户提供一些额外的信息，当我们的程序处理多个应用程序时也是如此。这些窗口由窗口管理器直接组织和管理，不需要每次都有任何父窗口与其关联。

语法：

```
toplevel = Toplevel(root, bg, fg, bd, height, width, font, ..)
```

可选参数描述如下。

- root：根窗口（可选）。
- bg：背景色。
- fg：前景色。
- bd：边框。
- height：小部件的高度。
- width：小部件的宽度。
- font：文本的字体类型。
- cursor：出现在小部件上的光标，可以是箭头、点等。

常用方法描述如下。

- iconify()：将窗口变成图标。
- deionify()：将图标转回窗口。
- state()：返回窗口的当前状态。
- withdraw()：将窗口从屏幕上删除。
- title()：定义窗口的标题。
- frame()：返回特定于系统的窗口标识符。

示例代码如下：

```
from tkinter import *

root = Tk()
root.geometry("200x300")
root.title("main")

l = Label(root, text = "This is root window")

top = Toplevel()
top.geometry("180x100")
top.title("toplevel")
l2 = Label(top, text = "This is toplevel window")

l.pack()
l2.pack()

top.mainloop()
```

Scrollbar 小部件用于向下滚动内容。这个小部件提供了一个滑动控制器，用于实现垂直滚动的小部件，如 Listbox、Text 和 Canvas。

下面给出了使用 Scrollbar 小部件的语法。

```
w = Scrollbar(master, options)
```

参数描述如下。

- master：此参数用于表示父窗口。
- options：有许多可用的选项，它们可以用作用逗号分隔的键值对。

以下是可用于此小部件的常用选项。

- activebackground：此选项用于表示控件具有焦点时的背景色。
- bg：此选项用于表示小部件的背景色。
- bd：此选项用于表示小部件的边框宽度。
- command：此选项可以设置为与列表相关联的过程，每次移动滚动条时都可以调用该过程。
- cursor：在该选项中，鼠标指针将更改为该选项的光标类型，可以是箭头、点等。
- elementborderwidth：此选项用于表示箭头和滑块周围的边框宽度。默认值为 -1。
- highlightbackground：当小部件没有焦点时，此选项用于焦点高光颜色。
- highlightcolor：当小部件具有焦点时，这是焦点高光颜色。
- highlightthickness：此选项用于表示焦点高光的厚度。

- jump：此选项用于控制滚动跳转的行为。如果设置为 1，则在用户释放鼠标按钮时调用回调。
- orient：根据滚动条的方向，此选项可以设置为 HORIZONTAL（水平）或 VERTICAL（垂直）。
- repeatdelay：此选项告诉在滑块开始向该方向重复移动之前，按钮被按下的持续时间。默认值为 300 毫秒。
- repeatinterval：一旦滑块在某方向的持续移动开始，该值决定了相邻两次移动动作的时间间隔。
- takefocus：可以使用 Tab 键将焦点切换到 scrollbar。如果 takefocus=0，那么将关闭该功能。
- troughcolor：此选项用于表示槽的颜色。
- width：此选项用于表示滚动条的宽度。

此小部件中使用的方法如下。

- get()：返回描述滑块当前位置的两个数字 (a, b)。a 值分别为水平和垂直滚动条提供滑块左边缘或上边缘的位置；b 值给出右边缘或下边缘的位置。
- set (first, last)：设置滚动条的滑块的位置。

示例代码如下：

```
from tkinter import *

root = Tk()
root.geometry("150x200")

w = Label(root, text ='GeeksForGeeks',
          font = "50")

w.pack()

scroll_bar = Scrollbar(root)

scroll_bar.pack( side = RIGHT,
                 fill = Y )

mylist = Listbox(root,
                 yscrollcommand = scroll_bar.set )

for line in range(1, 26):
    mylist.insert(END, "Geeks " + str(line))
```

```
mylist.pack( side = LEFT, fill = BOTH )

scroll_bar.config( command = mylist.yview )

root.mainloop()
```

LabelFrame 是 Tkinter 中 Label 和 Frame 小部件的组合。默认情况下，LabelFrame 会在其子组件的周围绘制一个边框以及一个标题。

以下是在 Tkinter 中创建 LabelFrame 的语法。

```
Labelframe_tk = LabelFrame ( windows, features )
```

LabelFrame 的特征和属性描述如下。

- bg：显示小部件的背景色。
- bd：显示边框的宽度。
- cursor：在该选项中，鼠标指针将更改为该选项的光标类型，可以是箭头、点等。
- fg：确定用于小部件的字体的前景色。
- font：确定用于小部件的字体类型。
- height：确定小部件的高度。
- labelAnchor：指定文本在小部件中的位置。
- labelwidget：指定用于标识标签的小部件。如果未定义值，则使用 Text 作为默认值。
- highlightbackground：单击时显示文本小部件背景的高亮颜色。
- highlightcolor：显示单击标签框架小部件时的高亮颜色。
- highlightthickness：指定焦点中高亮的厚度。
- padx：在水平方向上添加填充。
- pady：在垂直方向添加填充。
- relief：显示不同类型的边框。默认情况下，它有一个 FLAT 边框。
- text：指定包含标签文本的字符串。
- width：指定小部件的宽度。

可以创建简单的 LabelFrame 小部件，如下所示：

```
from tkinter import *
screen = Tk()
screen.geometry('300x300')

labelframe_tk = LabelFrame(screen, text="LabelFrame Title")
labelframe_tk.pack(fill="both", expand="yes")

inside = Label(labelframe_tk, text="Add whatever you like")
```

```
inside.pack()

screen.mainloop()
```

在每个应用程序中,我们都需要显示一些消息,如"要关闭吗"或显示任何警告或其他信息,为此,Tkinter 提供了一个 messagebox 库。通过使用 messagebox 库,我们可以消息框的形式显示一些信息、错误、警告、取消等。它有一个不同的消息框,用于不同的目的。

(1) showinfo()——显示一些重要信息。

(2) showwarning()——显示某种类型的警告。

(3) showerror()——显示一些错误消息。

(4) askquestion()——显示带有两个选项"是"或"否"的对话框。

(5) askokcancel()——显示一个对话框,询问两个选项"是"或"取消"。

(6) askretrycancel()——显示一个对话框,询问两个选项"重试"或"取消"。

(7) askyesnocancel()——显示一个对话框,询问三个选项"是""否"或"取消"。

MessageBox 函数的语法:

```
messagebox.name_of_function(Title, Message, [, options])
```

(1) name_of_function——要使用的函数名。

(2) Title——消息框的标题。

(3) Message——要在对话框中显示的消息。

(4) Options——配置选项。

示例代码如下:

```
from tkinter import *
from tkinter import messagebox

root = Tk()
root.geometry("300x200")

w = Label(root, text ='GeeksForGeeks', font = "50")
w.pack()

messagebox.showinfo("showinfo", "Information")

messagebox.showwarning("showwarning", "Warning")

messagebox.showerror("showerror", "Error")
```

```
messagebox.askquestion("askquestion", "Are you sure?")

messagebox.askokcancel("askokcancel", "Want to continue?")

messagebox.askyesno("askyesno", "Find the value?")

messagebox.askretrycancel("askretrycancel", "Try again?")

root.mainloop()
```

place 几何管理器允许以绝对值或相对于另一个窗口显式设置窗口的位置和大小。可以通过 place() 方法访问位置管理器，该方法适用于所有标准小部件。对于普通的窗口和对话框布局，使用 place() 通常不是一个好主意，让事情按应有的方式运转，实在是太难了。为此，请使用 pack() 或 grid() 管理器。语法：

```
widget.place(relx = val, rely = val, anchor = VAL)
```

示例代码如下：

```
# importing tkinter module
from tkinter import * from tkinter.ttk import *

# creating Tk window
master = Tk()

# setting geometry of tk window
master.geometry("200x200")

# button widget
b1 = Button(master, text = "Click me !")
b1.place(relx = 1, x =-2, y = 2, anchor = NE)

# label widget
l = Label(master, text = "I'm a Label")
l.place(anchor = NW)

# button widget
b2 = Button(master, text = "GFG")
b2.place(relx = 0.5, rely = 0.5, anchor = CENTER)

# infinite loop which is required to
# run tkinter program infinitely
# until an interrupt occurs
mainloop()
```

当使用 pack() 或 grid() 管理器时，很容易将两个不同的小部件彼此分开，但将其中一个放到另一个里面有点困难，但这可以通过 place() 方法轻松实现。在 place() 方法中，我们可以使用 In_ 选项将一个小部件放在另一个内，示例代码如下：

```
# importing tkinter module
from tkinter import * from tkinter.ttk import *

# creating tk window
master = Tk()

# setting geometry of tk window
master.geometry("200x200")

# button widget
b2 = Button(master, text = "GFG")
b2.pack(fill = X, expand = True, ipady = 10)

# button widget
b1 = Button(master, text = "Click me !")

# this is where b1 is placed inside b2 with in_ option
b1.place(in_= b2, relx = 0.5, rely = 0.5, anchor = CENTER)

# label widget
l = Label(master, text = "I'm a Label")
l.place(anchor = NW)

# infinite loop which is required to
# run tkinter program infinitely
# until an interrupt occurs
mainloop()
```

grid 几何管理器将小部件放在二维表中。主窗口小部件被分成许多行和列，结果表中的每个"单元格"都可以容纳一个窗口小部件。grid 几何管理器是 Tkinter 中最灵活的几何管理器。示例代码如下：

```
# import tkinter module
from tkinter import * from tkinter.ttk import *

# creating main tkinter window/toplevel
master = Tk()

# this will create a label widget
l1 = Label(master, text = "First:")
```

```
l2 = Label(master, text = "Second:")

# grid method to arrange labels in respective
# rows and columns as specified
l1.grid(row = 0, column = 0, sticky = W, pady = 2)
l2.grid(row = 1, column = 0, sticky = W, pady = 2)

# entry widgets, used to take entry from user
e1 = Entry(master)
e2 = Entry(master)

# this will arrange entry widgets
e1.grid(row = 0, column = 1, pady = 2)
e2.grid(row = 1, column = 1, pady = 2)

# infinite loop which can be terminated by keyboard
# or mouse interrupt
mainloop()
```

使用 pack 命令,我们可以声明小部件彼此之间的位置。pack 命令负责详细信息,这是最容易实现的布局管理器。对于开发简单或小型的 GUI 应用程序,最好使用 pack 几何管理器。示例代码如下:

```
# geometry manager - pack
from tkinter import *

root = Tk()

w = Label(root, text="Red Zone", bg="red", fg="white")
w.pack()
w = Label(root, text="Green Glossy", bg="light green", fg="white")
w.pack()
w = Label(root, text="Yellow Yuga", bg="yellow", fg="red")
w.pack()

root.geometry("250x140")
root.mainloop()
```

可以与 pack 几何管理器一起使用的属性描述如下。

(1) Fill:用于填充整个区域(使用背景色),即水平(x)或垂直(y)。

(2) Padding:用于为元素提供相对于 x 轴或 y 轴的填充。

(3) Side:用于将元素放置在特定的边上,它可以是右侧或左侧。

绑定函数用于处理事件。我们可以将 Python 的函数和方法绑定到事件,也可以将这些函数绑定到任何特定的小部件。如果要绑定小部件的事件，请在该小部件上调用 .bind() 方法。绑定小部件事件的语法如下：

```
widget.bind(event, event handler)
```

参数描述如下。

- event——由用户引起的可能反映更改的事件。
- event handler——应用程序中在事件发生时调用的函数。

以下是如何将事件绑定到小部件的特定实例的示例。

```python
import tkinter as tk
class Display:
    def __init__(self):
        self.root = tk.Tk()
        self.entry1 = tk.Entry(self.root)
        self.entry1.bind("<KeyPress>", self.onKeyPress)
        self.entry1.pack()
        self.root.mainloop()

    def onKeyPress(self, event):
        print("Key has been Pressed.")
display = Display()
```

7.2 网络爬虫图形用户界面

带用户界面的网络爬虫抓取输入网址的源代码，并在新窗口中显示出来。代码如下：

```python
from tkinter import *
import tkinter as tk
from tkinter import messagebox as ms
from PIL import ImageTk, Image
from bs4 import BeautifulSoup
import urllib

def Scrape(arg=None):

    # Checking for blank url
    if url_entry.get() == '':
        ms.showerror('Oops', 'Enter A Valid URL !!!')

    else:
        try:
```

第7章 开发网络爬虫用户界面

```python
        ''' Scraping Method Start'''
        # Giving url
        url = url_entry.get()

        # Reading all content
        content = urllib.request.urlopen(url).read()

        # Passing the content to function
        soup = BeautifulSoup(content, features="lxml")

        # Storing html in one variable
        info = soup.prettify()
        '''Scrape Method End'''

        '''Window Settings Start'''
        # Creating New Window
        root = tk.Toplevel()

        # Creating Title
        root.title('Thank You For Using Our Service !!!!')

        # Creating title icon
        root.iconbitmap('img/logo.ico')

        # Locking the window size
        root.resizable(width=False, height=False)

        ''' Window Setting End'''

        # Adding scrollbar to the window
        scrollbar = Scrollbar(root)
        scrollbar.pack(side=RIGHT, fill=Y)

        # Using text widget to show scraped content
        text = Text(root, yscrollcommand=scrollbar.set, wrap = WORD)
        text.insert(INSERT, info)
        text.pack()

        # Scroll bar settings
        scrollbar.config(command=text.yview)

    except ValueError:
        ms.showerror('Error', 'Enter A Valid URL !!!')

''' Window Setting Start '''
```

```python
# Creating Widget
crawler = tk.Tk()

# Creating size of window
crawler.geometry('500x500')

# Locking the window size
crawler.resizable(width=False, height=False)

# Creating Title
crawler.title('Web Scraper for HTML & XML')

# Creating title icon
crawler.iconbitmap('img/logo.ico')
''' Window Setting End '''

# Top Frame
top_frame = Label(crawler, text='WEB CRAWLER',font = ('Cosmic', 25, 'bold'), bg='#C70039', fg='white', relief='groove',padx=500, pady=30, bd='5')
top_frame.pack(side='top')

''' Background Image Start'''
# Sizing Image
canvas = Canvas(crawler, width=500, height=500)

# Opening Image
image = ImageTk.PhotoImage(Image.open('img/bg6.jpg'))

#Positioning Image
canvas.create_image(0,0, anchor=NW, image=image)
canvas.pack()
'''Background Image End'''

# Creating Frame
frame = LabelFrame(crawler, padx=30, pady=40, bg='white', bd='5', relief='groove')
frame.place(relx = 0.5, rely = 0.5, anchor = CENTER)

# Label
url_add = tk.Label(frame, text = 'Enter a URL or Web Address',font=('Arial',10,'bold'),bg='white', fg='green').pack()

# Entry or Input
url_entry = tk.Entry(frame, font=('calibre',10,'normal'), justify = 'center', bg='#FBB13C', width='30')
```

```
# Returning value to the function
url_entry.bind('<Return>', Scrape)

# Setting focus for input
url_entry.focus_set()

# Placing the button
url_entry.pack()

# Label for seperating Buttons
label = Label(frame, bg='white').pack()

# Creating Submit button and positioning it
crawl = tk.Button(frame, text = "Scrape", width="10", bd = '3', command = Scrape,
font = ('Times', 12, 'bold'), bg='#7268A6',relief='groove', justify = 'center',
pady='5').pack()

# Creating window only once
crawler.mainloop()
```

7.3 本章小结

本章介绍了使用 Tkinter 模块实现网络爬虫图形用户界面的方法。

我们迄今为止所做的大多数程序都是基于文本的编程,但许多应用程序需要 GUI (Graphical User Interface,图形用户界面)。

Python 提供了标准库 Tkinter,用于为基于桌面的应用程序创建图形用户界面。Tkinter 有一个简单的语法,以及三个几何管理器,即 grid、place 和 pack。

第 8 章　案例分析

本章首先介绍影视采集器的案例,随后介绍抓取搜索引擎结果的暗网爬虫。

8.1　影视采集器

抓取影视网站 http://kusonime.com/。使用 Scrapy 实现这个爬虫。

创建一个名为 webcrawler 的 Scrapy 项目:

```
scrapy startproject webcrawler
```

创建 kusonime 爬虫:

```
cd webcrawler
scrapy genspider kusonime kusonime.com
```

实现爬虫的 kusonime.py 文件内容如下:

```
import scrapy


class KusonimeSpider(scrapy.Spider):
    name = 'kusonime'
    allowed_domains = ['kusonime.com']
    start_urls = ['http://kusonime.com/']

    def parse(self, response):
        for i in response.css(".episodeye a::attr(href)"):
            yield scrapy.Request(url=i.get(), callback=self.parse_content)

        next_page = response.css("link[rel='next']::attr(href)")
        if next_page:
            yield scrapy.Request(url=next_page.get())

    def parse_content(self, response):
        item = {
            "title": (
```

```
                response.css(".clear ~ p strong::text").get() or
                response.css(".wp-post-image::attr(title)").get()
            ).strip(),
            "url": response.url,
            "genre": response.css("a[rel='tag']::text").extract(),
            "thumbnail": response.css(".wp-post-image::attr(src)").get()
        }

        for info in response.css(".info p"):
            data = info.css("::text").getall()
            if data[0].strip() == "Genre":
                continue

            if len(data) > 2:
                k, v = data[0], data[-1]
            else:
                k, v = data
            item[k.strip()] = v.strip(": ")
        item["sinopsis"] = response.css(".clear ~ p ::text").get().strip()

        downloads = []
        for ddl in response.css(".smokeddl"):
            name = ddl.css(".smokettl::text").get()
            if not name:
                continue

            data = []
            for smokeurl in ddl.css(".smokeurl"):
                data.append({
                    "desc": smokeurl.css("strong::text").get(),
                    "url": smokeurl.css("a::attr(href)").extract()
                })
            downloads.append({
                "name": name,
                "link": data})
        item["download_data"] = downloads
        yield item
```

使用 SQLAlchemy 将数据存入 SQLite 数据库：

```
import sqlalchemy as sa
import warnings
from sqlalchemy import exc as sa_exc
from sqlalchemy.orm import sessionmaker
from sqlalchemy.orm import registry
```

```python
from collections import defaultdict
from urllib.parse import unquote
from pathlib import Path
import re
import attr
import sys
import copy
import logging
import difflib
import json

from scrapy.utils.project import get_project_settings

settings = get_project_settings()
db_name = settings.get("DATABASE_NAME", "database")

db_dir = Path(__file__).parent.joinpath("../database")
db_dir.mkdir(exist_ok=True)
db_path = db_dir.joinpath(db_name + ".sqlite")

warnings.simplefilter("ignore", category=sa_exc.SAWarning)


class database:
    def __init__(self):
        self.engine = sa.create_engine("sqlite:///%s" % db_path)
        self.mapper = registry()
        self.mapper.metadata.bind = self.engine
        Session = sessionmaker(self.engine)
        self.session = Session()

        self.columns = defaultdict(dict)
        self.unique_keys = defaultdict(lambda: None)
        self.table = None

    def create_new_table(self, dbname, columns, metadata):
        wrapper = attr.s(type("wrapper", (), metadata))
        wrapper = type(dbname.title(), (wrapper,), {
            "__table__": sa.Table(
                dbname, self.mapper.metadata, *columns, extend_existing=True)})
        return self.mapper.mapped(wrapper)

    def init_table(self):
        table, inspector = {}, sa.inspect(self.engine)
```

```python
        for dbname in inspector.get_table_names():
            columns, metadata = [], {"id": attr.ib(init=False)}
            for column in inspector.get_columns(dbname):
                name = column["name"]
                column_type = column.pop("type")
                column["type_"] = column_type

                sa_col = sa.Column(**column)
                columns.append(sa_col)
                self.columns[dbname][name] = sa_col

                if name != "id" or self.unique_keys[dbname] == "id":
                    metadata[name] = attr.ib(default=column_type.python_type())
            table[dbname] = self.create_new_table(dbname, columns, metadata)
        self.table = table

    def get_column_type(self, v):
        if isinstance(v, bool):
            t = sa.Boolean
        elif isinstance(v, int) or (isinstance(v, str) and v.isdigit()):
            t = sa.Integer
        elif isinstance(v, (str, dict, list)):
            t = sa.String
        else:
            raise ValueError(f"unknown value type {v!r}")
        return t

    def safe_name(self, name):
        return re.sub(r"\s+", "_", name.strip()).lower().strip()

    def create_table_from_data(self, dbname, data_dict):
        columns, metadata = [
            sa.Column("id", sa.Integer(), primary_key=True)], {
            "id": attr.ib(init=False)}
        for k, v in data_dict.items():
            name = self.safe_name(k)
            v_type = self.get_column_type(v)

            if name != "id" or self.unique_keys[dbname] == "id":
                metadata[name] = attr.ib(default=v_type().python_type())
            if name != "id":
                columns.append(sa.Column(name, v_type))

        # extend data
        for k, v in self.columns[dbname].items():
```

```python
            if k not in metadata:
                columns.append(v)
                metadata[k] = attr.ib(default=v.type.python_type())

        # update self.columns
        for col in columns:
            self.columns[dbname][self.safe_name(col.name)] = col

        self.table[dbname] = self.create_new_table(dbname, columns, metadata)

    def add_column(self, table_name, column):
        column_name = column.compile(dialect=self.engine.dialect)
        column_type = column.type.compile(self.engine.dialect)
        self.engine.execute('ALTER TABLE %s ADD COLUMN %s %s' %
                            (table_name, column_name, column_type))
        logging.info(f"add new column: '{column_name}' type {column_type!r}")

    def update_database(self):
        self.mapper.metadata.create_all(self.engine)

    def exists(self, name, filters: dict):
        name = self.safe_name(name)
        if not self.table.get(name):
            return False
        table = self.table[name]
        filters = [table.__table__.columns.get(
            key) == value for key, value in filters.items()]
        return self.session.query(table).filter(*filters).count() != 0

    def commit(self):
        self.session.commit()

    def rollback(self):
        self.session.rollback()

    def add(self, dbname, data_dict):
        dbname = self.safe_name(dbname)

        if not self.table.get(dbname):
            logging.info(f"create a new table: {dbname!r}")
            self.create_table_from_data(dbname, data_dict)
            self.update_database()

        table = self.table[dbname]
```

```python
            # rename similar key based on table column and filter none value
            columns, force_update = list(
                set(sorted(table.__table__.columns.keys()))), False
            for k, v in copy.copy(data_dict).items():
                sk = self.safe_name(k)
                v = data_dict.pop(k)

                if not v:
                    continue

                similar = difflib.get_close_matches(sk, columns)
                if len(similar) > 0:
                    sk = similar[0]
                if isinstance(v, (dict, list)):
                    data_dict[sk] = json.dumps(v, indent=2)
                else:
                    data_dict[sk] = v

                if sk not in columns:
                    force_update = True
                    v_type = self.get_column_type(v)
                    self.add_column(dbname, sa.Column(sk, v_type))

            if force_update:
                self.create_table_from_data(dbname, data_dict)
                table = self.table[dbname]

            self.session.add(table(**data_dict))


class ProcessPipeline:
    def __init__(self):
        self.db = None if "-o" in sys.argv or "--output" in sys.argv else database()
        self.names = {}

    def parse_dbname(self, spider):
        clsname = spider.__class__.__name__
        return self.db.safe_name(clsname[0] + re.sub(r"[A-Z]", lambda x: "_" + x[0], clsname[1:-6]))

    def process_item(self, item, spider):
        if not item:
            return
```

```
            unique_key = getattr(spider, "unique_key", "title")
            output_keys = getattr(spider, "outputs", [unique_key])
            out = ", ".join(f"{item.get(key)!r}" for key in output_keys)

            if not self.db:
                logging.info(f"{out} crawled")
                return item

            if not self.names.get(spider.name):
                self.names[spider.name] = self.parse_dbname(spider)
            name = self.names[spider.name]

            self.db.unique_keys[name] = unique_key
            if self.db.table is None:
                self.db.init_table()

            if self.db.exists(name, {unique_key: item[unique_key]}):
                logging.error(f"{out}: already exists!")
            else:
                try:
                    if item.get('url'):
                        item["url"] = unquote(item["url"])
                    self.db.add(name, item)
                    self.db.commit()
                    logging.info(f"{out}: added to database")
                except Exception as e:
                    self.db.rollback()
                    logging.error(
                        f"{out}: failed added to database\n{e}")
            return item
```

8.2 暗网爬虫

在大海中，很多鱼深藏在水下。有的页面就像是深不可测的大海，很多有用的信息隐藏在网页中。

暗网的表现形式一般是：前台是一个表单来获取，提交后返回一个列表形式的搜索结果页，它们是由暗网后台数据库动态产生的。搜索引擎本身也可以看作一个暗网。搜索结果页包含了指向详细内容页的链接。搜索引擎中的内容有挖掘价值。

search_engines(https://github.com/tasos-py/Search-Engines-Scraper) 是一个 Python 库，用于查询 Google、Bing、Yahoo 和其他搜索引擎并从多个搜索引擎的结果页面中收集结果。

为了易于为 search_engines 添加新的搜索引擎，可以通过在 search_engines/engines/ 中创建一个新类来添加新引擎，并将其添加到 search_engines/engines/__init__.py 中的 search_engines_dict 字典中。新类应该为 SearchEngine 类的子类，并覆盖以下方法：_selectors、_first_page、_next_page。

为了安装这个库，需要运行设置文件：

```
python setup.py install
```

用法如下：

```
from search_engines import Bing

engine = Bing()
results = engine.search("java")
links = results.links()

print(links)
```

8.3 本章小结

本章介绍了影视采集器和暗网爬虫的案例。